BUSINESS
INTELLIGENCE FOR
TELECOMMUNICATIONS

OTHER TELECOMMUNICATIONS BOOKS FROM AUERBACH

Architecting the Telecommunication Evolution: Toward Converged Network Services
Vijay K. Gurbani and Xian-He Sun
ISBN: 0-8493-9567-4

Business Strategies for the Next-Generation Network
Nigel Seel
ISBN: 0-8493-8035-9

Chaos Applications in Telecommunications
Peter Stavroulakis
ISBN: 0-8493-3832-8

Context-Aware Pervasive Systems: Architectures for a New Breed of Applications
Seng Loke
ISBN: 0-8493-7255-0

Fundamentals of DSL Technology
Philip Golden, Herve Dedieu, Krista S Jacobsen
ISBN: 0-8493-1913-7

Introduction to Mobile Communications: Technology, Services, Markets
Tony Wakefield
ISBN: 1-4200-4653-5

IP Multimedia Subsystem: Service Infrastructure to Converge NGN, 3G and the Internet
Rebecca Copeland
ISBN: 0-8493-9250-0

MPLS for Metropolitan Area Networks
Nam-Kee Tan
ISBN: 0-8493-2212-X

Performance Modeling and Analysis of Bluetooth Networks: Polling, Scheduling, and Traffic Control
Jelena Misic and Vojislav B Misic
ISBN: 0-8493-3157-9

A Practical Guide to Content Delivery Networks
Gilbert Held
ISBN: 0-8493-3649-X

Resource, Mobility, and Security Management in Wireless Networks and Mobile Communications
Yan Zhang, Honglin Hu, and Masayuki Fujise
ISBN: 0-8493-8036-7

Security in Distributed, Grid, Mobile, and Pervasive Computing
Yang Xiao
ISBN: 0-8493-7921-0

TCP Performance over UMTS-HSDPA Systems
Mohamad Assaad and Djamal Zeghlache
ISBN: 0-8493-6838-3

Testing Integrated QoS of VoIP: Packets to Perceptual Voice Quality
Vlatko Lipovac
ISBN: 0-8493-3521-3

The Handbook of Mobile Middleware
Paolo Bellavista and Antonio Corradi
ISBN: 0-8493-3833-6

Traffic Management in IP-Based Communications
Trinh Anh Tuan
ISBN: 0-8493-9577-1

Understanding Broadband over Power Line
Gilbert Held
ISBN: 0-8493-9846-0

Understanding IPTV
Gilbert Held
ISBN: 0-8493-7415-4

WiMAX: A Wireless Technology Revolution
G.S.V. Radha Krishna Rao, G. Radhamani
ISBN: 0-8493-7059-0

WiMAX: Taking Wireless to the MAX
Deepak Pareek
ISBN: 0-8493-7186-4

Wireless Mesh Networking: Architectures, Protocols and Standards
Yan Zhang, Jijun Luo and Honglin HU
ISBN: 0-8493-7399-9

Wireless Mesh Networks
Gilbert Held
ISBN: 0-8493-2960-4

AUERBACH PUBLICATIONS

www.auerbach-publications.com
To Order Call: 1-800-272-7737 • Fax: 1-800-374-3401
E-mail: orders@crcpress.com

BUSINESS INTELLIGENCE FOR TELECOMMUNICATIONS

Deepak Pareek

Auerbach Publications
Taylor & Francis Group
Boca Raton   New York

Auerbach Publications is an imprint of the
Taylor & Francis Group, an informa business

Auerbach Publications
Taylor & Francis Group
6000 Broken Sound Parkway NW, Suite 300
Boca Raton, FL 33487-2742

© 2007 by Taylor & Francis Group, LLC
Auerbach is an imprint of Taylor & Francis Group, an Informa business

No claim to original U.S. Government works
Printed in the United States of America on acid-free paper
10 9 8 7 6 5 4 3 2 1

International Standard Book Number-10: 0-8493-8792-2 (Hardcover)
International Standard Book Number-13: 978-0-8493-8792-0 (Hardcover)

Library of Congress Cataloging-in-Publication Data

Pareek, Deepak
    Business intelligence for telecommunications / Deepak Pareek.
       p. cm.
    Includes bibliographical references and index.
    ISBN 0-8493-8792-2 (alk. paper)
    1. Business intelligence--Management. 2. Business intelligence--Data processing. 3. Telecommunication--Management. I. Title.

HD38.7.P373 2006
384.068'4--dc22   2006048258

Visit the Taylor & Francis Web site at
http://www.taylorandfrancis.com

and the Auerbach Web site at
http://www.auerbach-publications.com

TABLE OF CONTENTS

PREFACE

What is it that drives businesses to success?

Although business intelligence systems are widely used in industry, research about them is limited. The objective of this tutorial is to describe the state-of-the-art of business intelligence and to suggest potential IS research topics.

Most analysts when looking into the future trying to identify technologies that will change the way businesses will be conducted invariably come out with Business Intelligence (BI) at the top of the list. No doubt today BI is one of the technology areas that are hyped, but in the case of BI it would be appropriate to add the suffix "under" to the word "hyped."

Organizations that need to gain more efficiency and manage or reduce costs are looking to BI to address their requirements. Business intelligence provides timely and accurate information to better understand your business and to make more informed, real-time business decisions.

Today's business managers need factual information at their fingertips to anticipate changes and make proactive decisions. Business intelligence and data management help business managers make better decisions by converting data into useful information. BI is important in helping companies stay ahead of the competition by providing the means for quicker, more accurate, and more informed decision making. The most agile BI products and services are not confined by industry classification and can create an infinite number of possible applications and of value-increasing possibilities for any business department or combination of departments.

This book, which contains eleven chapters, is an excellent resource for those interested in understanding the techno-business aspect of business intelligence and the impact of business intelligence developments on the telecommunications industry.

1

BUSINESS INTELLIGENCE: INTRODUCTION

Business intelligence is a discipline of developing information that is conclusive, fact-based, and actionable. Business intelligence gives enterprises the ability to discover and utilize information they already own, and turn it into the knowledge that directly affects corporate performance.

Information is the hottest commodity in business today. The information universe is transforming. Today succeeding in business depends on how well organizations know their customers, how well they understand their business processes, and how effectively they run their operations. And having that kind of far-reaching insight depends on information: information that is accessed, integrated, and distributed in a meaningful fashion.

The technological developments of this decade and the past century have created an ever-expanding universe of information, from pure content creation to the endless statistics created by every individual's interaction with the new technological environment. This growth in the ability to create and distribute information along with the facility to monitor its use offers enormous cultural, educational, and financial benefits for those who know how to harness it. However, the same growth has led to an inevitable dilution in both the quality and the applicability of that information to any particular individual unable to utilize knowledge of information.

Monitoring and identifying relevant information, and assimilating, managing, and responding to it, are placing increasing time demands on the individual. In an ever-expanding universe of information, information itself is no longer scarce but knowledge is. Many people have reached a point of "information overload"; they do not have the time to find the valuable information themselves and convert it into knowledge.

Table 1.1 Goals of the Modern Enterprise

Transparency
  To management/command chain: Business intelligence
  To the citizen
  To other agencies
  Of events: Business Activity Monitoring (BAM)

Integration
  Automate processes to reduce cost and increase reliability
  Integrate with other commands/agencies
  Adoption of industry standards and Web service technology
  Agility to change and customize

IT Efficiency
  Consolidation of redundant systems
  Productivity of developers

Similarly for enterprises, systems are delivering and receiving information on a continuous basis: client services deliver data and track usage; internal services aggregate information; and systems monitor technical activities around the clock (see Table 1.1). The real challenge is to generate information using this data and harness this information for creating knowledge, hence real competitive advantage, and for economic benefit.

OVERVIEW

Today's enterprises are judged not only on the quality of their products and services, but also on how well they share information with customers, employees, and business partners. The more widely available information is throughout the enterprise, the more valuable it becomes. When a marketing department has accurate data about the installed base of products and services, it is better able to develop targeted promotions. When customers can easily check if an item is in stock, they are more likely to make a purchase. When senior executives have instant access to trend data, they can turn on a dime in the most profitable direction for the firm.

The problem is that all the information people rely on is not in one place. Most organizations have myriad systems, each with its own data sources and presentation mechanisms. This makes maintaining complete, up-to-date information across many departments and business units extremely difficult. The more integrated an enterprise becomes, the easier it is for everyone to get the information they need, so they are empowered to make their best decisions.

That means organizations need ready and useful access to the intelligence that resides in their operational data. Many attempts have been made to give people throughout organizations access to this information from submitting report requests to IT departments, to extracting and downloading operational data into PCs for use with spreadsheets, or to executive information systems. Unfortunately, past solutions have exhibited many problems, including significant delays in the delivery of timely information, inconsistencies of data, and the non-flexibility of these procedures to change as information needs evolve. And although many organizations have made large investments in operational systems (point-of-sale, order entry, manufacturing, work schedule, purchasing, etc.), these systems typically do not allow people to easily query, report, and analyze information.

What is required is a system that enables people to examine relationships and trends in all of their organization's data, no matter where it resides. An intelligent system that is designed to store all business data in integrated, subject-oriented databases that also can provide a historical perspective on the information is the key to ad hoc analysis and decision making.

INTELLIGENCE NOT JUST INFORMATION

Intelligence is the ability to learn, to understand, or to deal with new or trying situations; the skilled use of reason; or the ability to apply knowledge to manipulate one's environment or to think abstractly. Information fuels the new economy and plays an essential role in developing and maintaining a sustainable competitive advantage. The demands on a business today—increased global competition, lower barriers to entry, lower profit margins—are creating an ever-increasing need for access to data. Keys to success in such a trying business environment are:

How easily and consistently can organizations share financial results with managers?

How effortless is it for customers to learn about the organization's products?

How often do sales and support teams work in concert when a new product is released or when an old one is updated?

How quickly can organizations take the pulse of the overall organization?

The ability to get the right information to the right people at the right time is therefore more important than ever; however, the sheer volume of available data makes such a proposition more challenging than ever. Organizations that are the most successful at collecting, evaluating, and applying information are consistently the leaders in their respective industries. The ability to act faster and more effectively than the competition can be the defining advantage in today's marketplace and the means for successfully managing customer relationships in the long run.

Furthermore, increasing regulatory rigidity and enhanced competitive pressures are making present-day enterprises need analytical "answers" in real-time so they can plan more accurately and react more quickly to changing conditions and they also need to perform analytics across their organizations. All this leads to a complex scenario further compounded by the size and global footprint of today's enterprise.

As organizations look to address these challenges while they follow the dream to become leaders in a knowledge economy, with limited stand-alone two-dimensional tools, such as the spreadsheet, such complexity cannot be managed. There are several good solid business reasons for this assumption. A few of them are:

- Loss of opportunities due to unpreparedness
- Set goals not achieved regularly
- Continuous changing of business objectives
- Inability to scale the business
- Always reacting to events, and not in a timely manner
- Redoing or repeating processes, for example, rekeying data into different systems
- Conflicting customer treatment

It is easy to understand that these symptoms are all related to one another and are parts of a far more profound puzzle that can be solved at the enterprise level (see Figure 1.1).

Many organizations have already realized this, whereas others are going to realize it sooner or later. One step in the direction of making this complex

Figure 1.1   Blurring enterprise lines

scenario a strategic competitive advantage is proliferation of business intelligence among the enterprises. Ever since computers came on strong in the 1980s, enterprises have been trying to figure out ways to harness the power of all the data they created.

Intelligent Enterprise

Worldwide it is believed that information is the key to accomplish business goals and objectives and CEOs and CIOs across the globe feel that information is power and a valuable asset. Few executives deny that they want their enterprises to be more intelligent, meaning that they are able to react and plan by using information better.

However, it is a challenge to define and characterize what an intelligent enterprise looks like. Because different organizations use and process information differently and express different levels of maturity and evolution, most have different visions of an intelligent enterprise. However, all of them converge about end goals.

The ideal of an intelligent enterprise sounds like an easy concept to grasp. But how can organizations achieve it? The key to using information successfully is not how far the organization has evolved, but how it evolves. The method an enterprise uses to evolve determines if it can be considered an intelligent enterprise.

The intelligent enterprise shows sound judgment and rationality in planning a practical approach to delivering solutions that meet the long-term information needs of the organization. This does not mean that the intelligent enterprise has all the latest and greatest technology. It means that an intelligent enterprise exploits information to establish and maintain strategic vision. The intelligent enterprise attains strategic and tactical objectives through information usage to create Return On Investment (ROI) and sustainable competitive advantage. This can be done with simple or complex technology.

Organizations must recognize the benefits of maturing their intelligence capabilities proactively. Those that plan their evolution in a systematic and orderly manner typically gain over those that get pushed into change. The intelligent enterprise takes control of its own evolutionary path. It anticipates the need to grow, plans its next generation of information capabilities, and executes an information management plan that satisfies current and future needs (see Table 1.2).

As an organization moves along the evolutionary process, it expands its capability for delivering increased business value via information. Understanding how to evolve to an intelligent enterprise means the organization has to comprehend the steps along the way. Organizations need to recognize where they are and why, and then understand how to move to the next level. They need to agree on the level of evolution they

Table 1.2   Enterprise Business Needs and Associated Technologies

Need	Technology
Catalog data assets, track change, publish	Enterprise metadata repository
Common business language	Ontology model (including rules) for customer, items, orders, etc.
Attach objective business meaning to data	Mapping of data schemas to ontology
Manage information	Information management services—discovery, impact analysis, redundancy, classification
Integrate information	Information integration services— automatically infer and maintain data transformations and query
Virtual enterprise database	Federated query

need to achieve, establish priorities for improvement, and then implement pragmatic action plans for improving their information maturity so that it aligns with organizational goals and objectives.

Business Perspective of Intelligence

Initially business intelligence was considered a tool for getting information quickly and easily from the huge piles of data generated by business systems. But slowly and surely enterprises realized that business intelligence goes deeper than just data querying. Organizations that have successfully deployed either a department-based or enterprisewide business intelligence system not only have improved internal decision making within multiple business units or throughout the enterprise, but many also are able to extend this ready information access to customers, suppliers, and business partners.

Business intelligence is a set of concepts, methods, and processes to improve business decisions using information from multiple sources and applying experience and assumptions to develop an accurate understanding of business dynamics. Sometimes business intelligence refers to online decision making; it also refers to shrinking the information time window so that the intelligence is still useful to the decision maker when the decision time comes.

Businesses made an enormous investment in technology in the past couple of decades. In today's economy, spending has decreased and businesses are looking toward their technology investments and means to leverage these investments. However, although overall IT spending has decreased, within that spending, the portion spent on business intelligence and business intelligence-related technology has increased.

Numerous benchmark surveys of IT decision makers suggest that the majority of businesses will buy business intelligence tools in the coming quarters. A proof of the enhancing clout of business intelligence is that even in this tough economy, it is business intelligence along with enterprise business integration that is dominating IT investment. This growth trend can be attributed primarily to these three causes.

First, in the late 1990s enterprises spent millions to deploy customer relationship management, supply chain management, enterprise resource planning (ERP), and E-commerce software. These business applications have created reams of data about customers, suppliers, financial performance, and operating statistics. To turn this data into usable actionable information, enterprises are turning to business intelligence applications.

Second, enterprises increasingly see their data repositories as sources of immediate and sustainable competitive advantage. For some enterprises, this means identifying their most profitable customers and lavishing them with extras. For others, it is leveraging activity-based costing to fully understand profitability. For still others, this means managing their sales and distribution channels tightly to maximize revenue. Business intelligence applications empower managers and executives with the insight to manage their operations closely and to react quickly to changing conditions.

Finally, in today's economic conditions, businesses scrutinize every aspect of their operations to find new revenue or squeeze out additional cost savings. Business intelligence facilitates this by supplying decision-support information.

WHAT IS BUSINESS INTELLIGENCE?

Business intelligence is a catchphrase that was coined in the mid-1990s to describe taking data from its raw form and turning it into something usable on which business decisions can be based. It is an umbrella term that ties together other closely related data disciplines including data mining, statistical analysis, forecasting, and decision support.

Business intelligence is a business strategy aimed at understanding and anticipating the needs of an enterprise's current requirements. It is knowledge about the enterprise's customers, competitors, business partners, competitive environment, and its own internal operations that gives management of the enterprise the ability to make effective, important, and often strategic business decisions.

The advent of the Internet has made business intelligence even more critical because organizations have a unified platform for easily distributing information to a wider range of decision makers inside and outside the walls of their organization: employees, suppliers, partners, and customers.

As organizations move to implement new systems on the Web, they are gathering more and more data about customers, markets, products, and processes, all of which can contribute to greater insight and business acuity. The nearly infinite reach of the Internet has made external data sources become easily available that can further enhance an organization's decision making. With the wealth of information organizations have accumulated in all of their production databases and data warehouses, they have found the gold mine.

Business intelligence systems combine operational data with analytical tools to present complex and competitive information to planners and decision makers (see Figure 1.2). Their objective is to improve the timeliness and quality of the input to the decision process. Business intelligence is used to understand the capabilities available in the firm; the state of the art, trends, and future directions in the markets; the technologies and the regulatory environment in which the firm competes; and the actions of competitors and the implications of these actions.

The emergence of the data warehouse as a repository, the advances in data cleansing, the increased capabilities of hardware and software, and the emergence of the Web architecture all combine to create a richer business intelligence environment than was available previously.

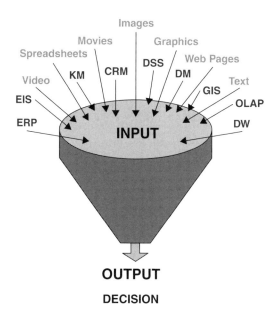

Figure 1.2   Business intelligence: generic model

Understanding Business Intelligence

Business intelligence means different things to different people, but essentially business intelligence encapsulates everything from operational reporting to data mining (see Figure 1.3 and Table 1.3). Business intelligence represents a broad category of applications and technologies for providing access to data to help enterprise users make better business decisions. Business intelligence includes:

- Decision support systems
- Forecasting
- Reporting data warehouse
- Data mart
- Data store
- Data mining
- Statistical analysis
- Extract, Transform, and Load (ETL)
- OnLine Analytical Processing (OLAP)
- Reporting
- Portal
- Ad hoc query

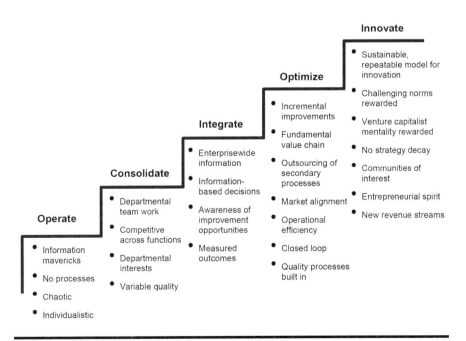

Figure 1.3   Information management evolution steps

Table 1.3   Business Intelligence Pitfalls

Incorrect data
  Missing data
Lack of sufficient data detail
  Insufficient security and privacy
Inflexible reports and queries
  Tools that are too difficult to use
Data not available timely enough
  Too difficult to add new data sources
Lack of user-oriented metadata
  Lack of system-oriented metadata
Inability to intermix ERP, CRM, and other data
  "Deskbound" analytics

The tools mentioned above that analyze data and information to help people acquire insight and knowledge are now much more accessible to a larger swath of employees, thanks to business intelligence. These business intelligence tools gather, manage, and analyze data to produce information that is distributed to people throughout the enterprise to improve strategic and tactical decisions, and at the same time condense the cycle between analysis and decision making as well as into real-time (see Table 1.4).

What Does Business Intelligence Do?

Business intelligence converts data into useful information and, through human analysis, into knowledge. Business intelligence is capable of performing

Table 1.4   Business Benefits

Early warning
  Business opportunities
Current and future threats
  Strategic decision making and plans
Alliances and acquisitions
  Major capital expenditures
New businesses, markets, and technology
  Competitive strategies and operations
Strategies directed at specific competitors
  Technology sourcing for product development
Supporting marketing, sales, and manufacturing
  Counterintelligence and security
Knowing what the competitors know about us
  Protecting our information and intellectual property

important tasks such as creating forecasts based on historical data, past and current performance, and estimates of the direction in which the future will go; "what if" analysis of the impacts of changes and alternative scenarios; and ad hoc access to the data to answer specific, non-routine questions and strategic insights to name a few. Business intelligence assists in strategic and operational decision making. The strategic uses of business intelligence are:

- Corporate performance management
- Optimizing customer relations, monitoring business activity, and traditional decision support
- Packaged stand-alone business intelligence applications for specific operations or strategies
- Management reporting of business intelligence

Business Intelligence Spectrum

Business intelligence provides exceptional functionality for a wide spectrum of business parameters (see Figure 1.4). It has a high applicability as it covers

- Different categories of business intelligence users
  - Report producers
  - Report consumers
- Different methods of delivery
  - Printed output, e-mail, Web, intranet, desktop applications
- Different types of reports
  - Operational, line of business reports
  - High-level analytic and forecasting reports
- Different types of access
  - Static, distributed reports
  - Ad hoc data access

Business Intelligence Audience

In the present complex business environment, issues such as increasing efficiencies, decreasing costs, and retaining customers all drive the importance of making more decisions, more often. This means more users in every level of the organization are responsible for analyzing more data than ever before, and the value of business intelligence is applicable from upper level management to the front line employee (see Figure 1.5). However, executives, managers/knowledge workers, and production/clerical staff all have very unique metrics and information needs, hence the applications, usage, and utility of business intelligence tools are heavily based on a segmentation of the business intelligence

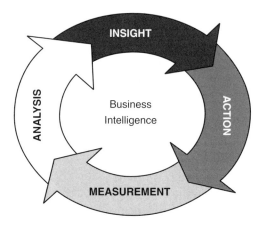

Figure 1.4   Business intelligence life cycle

user community. Different user segments will require different modes of delivery of business intelligence, ranging from basic static reporting to sophisticated analytic applications. Deploying the proper mix of tools and applications depending on end-users' needs is critical to achieving optimal benefits from business intelligence.

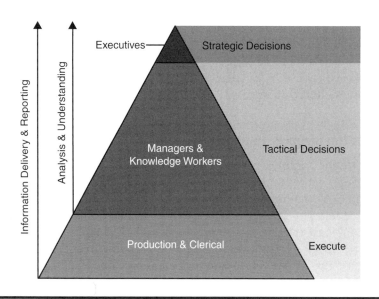

Figure 1.5   Decision-making pyramid

Table 1.5   Business Intelligence Hierarchy

Operational reporting
  What is happening right now?
Information analysis
  What happened yesterday, last week, last year?
Forecasting
  What is going to happen tomorrow, next week, next year?
Data mining
  How can I leverage the hidden patterns in my overload of data to uncover
  opportunities and challenges?

Executives

Increasingly more users at the executive level need not only to receive periodic summary information, but also more important, they need to understand the factors behind the data. Because of this need, the speed at which these systems can reveal insights must accelerate.

Executive decision makers are almost always on the leading edge of strategic decisions that have far-reaching impact on the rest of the organization, and depending on the nature of the organization, executives may also be involved to some degree in tactical decisions. Their work style is less about focus and more about monitoring a wide variety of factors that require constantly shifting gears to deal with the next problem or opportunity. Executive data usage and analysis needs are relatively simple and at the high-level end of the spectrum: data monitoring with the occasional need to look deeper quickly using tools such as time series analysis or guided root cause analysis. Executives need a business intelligence solution that is easy to use, nonintrusive, and customized to give them an "at a glance" overview of the top-level business Key Performance Indicators (KPIs). When more detailed analysis is necessary, a structured guide to help them research is highly useful. They do not have the time or the patience to learn to work with complex technology or to dig through mounds of data.

Managers, Supervisors, and Knowledge Workers

Users in these roles have responsibility for meeting certain business objectives within a specific time frame. In general, it is the managers and knowledge workers who invest the most time in decision making, as they typically make many more of the daily tactical decisions for the organization.

Business managers must make decisions about what course to take, what is working and what is not, and where to shift the focus of their teams. Knowledge workers analyze data for their own purposes, often to drive focus and trade-off decisions regarding spending allocations, projects with outside contractors/agencies, and so on. These two groups are most often responsible for tactical decisions and are held accountable for the successful execution of the corporate strategy. Their data usage and analysis modes cover the gamut from simple to complex and from reading and monitoring to advanced analysis. These users not only need to see critical information in an easy-to-digest format, but they also must have the ability to interrogate information and change the data requests in order to adapt to dynamic business situations and get answers to questions as conditions evolve. This group of tactical decision makers rarely has the luxury of high-level analyst support that is common to executives. Subsequently, although they have more focused needs, managers and knowledge workers typically have very little support in helping to digest the information and extract key insights on a timely basis.

Production and Clerical Staff

The role of production and clerical workers is execution. Similar to employees at all levels within the organization, these workers need information in order to perform their duties. However, as these people do not have traditional decision-making roles, the information they need is usually very narrow in scope (such as a copy of an invoice) and well understood ahead of time. Their data usage is in the realm of reading and perhaps some real-time monitoring of the activities they are assigned. Information needs for production and clerical staff is about fast access to predefined information and ease of use.

BUSINESS INTELLIGENCE HERITAGE

Although the term "business intelligence" is new, computer-based business intelligence systems go back, in one guise or other, for close to four decades. As computer hardware and software matured, with each new iteration, capabilities increased as enterprises grew ever more sophisticated in their computational and analytical needs. New innovative style products evolved based on sophisticated statistical and retrieval algorithms, initially described as decision support systems, then executive information systems, then knowledge management, and, more recently, business intelligence systems or corporate performance measurement.

Business Intelligence as a Decision Support System (DSS)

The developments associated with business intelligence go back to initiatives in 1985 to build a Decision Support System (DSS) that linked sales information and retail scanner data. Specialization and segmentation followed into complex data retrieval, data modeling, and, now, data mining, whereas business intelligence as a term replaced earlier decision support systems. The term business intelligence is a popularized umbrella term introduced by Howard Dresner of the Gartner Group in 1989.

Business intelligence describes a set of concepts and methods to improve business decision making by using fact-based support systems. Business intelligence is sometimes used interchangeably with briefing books, report and query tools, and executive information systems. Business intelligence systems are data-driven DSS.

Business intelligence finds its roots as a DSS and information systems researchers and technologists have built and investigated decision support systems for more than 40 years. The developments in DSS begin with building model-oriented DSS in the late 1960s.

Decision Support Systems (DSS) History

The history of decision support systems covers a relatively brief span of years, and the concepts and technologies are still evolving. In a technology field as diverse as DSS, history is not neat and linear. Different people have perceived the field from various vantage points and so they report different accounts of what happened and what was important.

Decision support systems evolved early in the era of distributed computing. Decision support pioneers include many academic researchers from programs at MIT, the University of Arizona, the University of Hawaii, the University of Minnesota, and Purdue University. The DSS pioneers created particular and distinct streams of technology development and research that serve as the foundation for much of today's work in DSS.

The DSS threads related to model-oriented DSS, expert systems, multidimensional analysis, query and reporting tools, OLAP, business intelligence, group DSS, and executive information systems are traced and interwoven as they appear to converge and diverge over the years. The history of such systems begins in about 1965 and it is important to start formalizing a record of the ideas, people, systems, and technologies involved in this important area of applied information technology.

Executive Information Systems (EIS) evolved from single-user model-driven decision support systems and improved relational database products. The first EIS used predefined information screens and were maintained by analysts for senior executives. Beginning in about 1990, data warehousing and OLAP began broadening the realm of EIS and defined a broader category of data-driven DSS.

Prior to 1965, it was very expensive to build large-scale information systems. At about this time, the development of the IBM System 360 and other more powerful mainframe systems made it more practical and cost-effective to develop Management Information Systems (MIS) in large enterprises. MIS focused on providing managers with structured periodic reports. Much of the information was from accounting and transaction systems.

In the late 1960s, a new type of information system became practical: model-oriented DSS or management decision systems. Around 1970 business journals started to publish articles on management decision systems, strategic planning systems, and decision support systems. By the late 1970s, a number of researchers and enterprises had developed interactive information systems that used data and models to help managers analyze semi-structured problems. These diverse systems were all called decision support systems.

From those early days, it was recognized that DSS could be designed to support decision makers at any level in an organization. DSS could support operations, financial management, and strategic decision making. A variety of models was used in DSS including optimization and simulation. Also, statistical packages were recognized as tools for building DSS. Artificial intelligence researchers began work on management and business expert systems in the early 1980s.

Financial planning systems became popular decision support tools. The idea was to create a "language" that would allow executives to build models without intermediaries. One major advantage that a planning language has over a spreadsheet is that the model is written using natural language and the model can be separated from the data. In the early 1980s, spreadsheets were also used for building model-driven DSS. Research related to using models and financial planning systems for decision support was encouraging, but certainly not uniformly positive.

Beginning in about 1990, DSS built using relational database technologies came to the forefront. At the same time a major technology shift occurred from mainframe-based DSS to client/server-based DSS. Some desktop OLAP tools were introduced during this time period. In 1992 to 1993, some vendors started recommending object-oriented technology for building "reusable" decision support capabilities. In 1994, many enterprises started to upgrade their network infrastructures. DBMS vendors understood and recognized that decision support was different from OLTP and started implementing real OLAP capabilities into their databases. Around 1993 the data warehouse and the EIS people found each other and the two niche technologies have been converging.

In 1995, data warehousing and the World Wide Web began to have an impact on practitioners and academics interested in decision support technologies. Web-based and Web-enabled DSS became feasible in the last few years of the past century. The Internet and Web have sped up developments in decision support and have provided a new means of capturing and applying knowledge.

EVOLUTION OF BUSINESS INTELLIGENCE

Historically, the first form of modern business intelligence solutions surfaced as information systems, dating back to the 1970s. Modern Business Intelligence (business intelligence) software originated with products released progressively from 1975 onward with initial products catering to financial and sales planning. As these systems matured and information from other functions within the organization funneled into them, the term "data warehouse" began surfacing. Spreadsheets became very popular from the 1980s on and these applications made the provision of both raw data and processed information more available.

Consequently, organizations began building applications and solutions that took advantage of this consolidated view of infomation. Transaction processing systems, for applications such as inventory management, order processing, costing, and the like, became widely available in corporations, creating new databases as a byproduct. As organizations began to evolve these solutions, the noun–data warehouse shifted to the verb–data warehousing to include both the physical structure and the applications sitting on top. Bundling both technology and business solutions together using such a common expression presented confusion to many business professionals.

The business application sitting on top of the enabling technology was clearly the most dominant of the two and, rightly so. Primarily because how individuals interact and use data warehouse solutions is the most important characteristic. How the solution is being used to solve real problems is essential. Better decisions and higher intelligence go hand in hand. Systems to report from the new data resources then made their first appearance. IBM's Structured Query Language (SQL) and Query by Example (QBE) retrieval packages were novel approaches to retrieval.

Business intelligence surfaced as an updated way to describe these solutions and enabling framework. This term quickly took hold. However, for many organizations the push to deploy business intelligence has been frustrating at best and at worst a costly and time-consuming failure. As business analytics evolved from basic period reporting to first-generation OLAP, more advanced ROLAP, and later to packaged metrics and data warehousing systems, enterprises struggled to balance the time and cost of implementation against their driving need for relevant and usable business information. Business intelligence was conceived to deliver focused and usable enterprise-level insights.

Today, the evolving definition for business intelligence can best be described as a framework that encompasses both solutions and enabling technology components designed to enhance the decision-making process within an organization. How one uses the framework is very important.

Does the framework allow business users to make the best decisions? Does the insight obtained from the business intelligence framework answer questions? Are the answers that organizations are getting the best? Today, many organizations are forced to make do with less. Not making the right decisions or addressing questions that surface around a given business problem with suboptimal answers can mean the difference between success and failure.

It is no longer sufficient for a business intelligence tool to access limited types of data. To fulfill the needs of a globally focused organization with disparate data sources, today's business intelligence solution must be able to access, integrate, and cross-reference new data, external data, and legacy data. Cross-referencing data in external public sources can increase the value of our own internal data. For example, an insurance organization could cross-reference its own data warehouse of insurance rates with the rates of competitors stored in external public sources, to locate areas and markets where they are most effective and those where they are not. This would allow it to customize its marketing programs and adjust its rates to be more competitive.

THE PRESENT

As organizations look to business intelligence solutions in order to address their critical business challenges, business intelligence software has become a multibillion dollar market. In the traditional world of business intelligence, most systems are "big ticket" investments that require the construction of massive and cumbersome data-mart infrastructures and the hiring and training of costly IT personnel.

Those expansive solutions are driven primarily by the revenue requirements of hardware and software enterprises, as opposed to the intelligence needs of corporate managers. Since the late 1990s, organizations have invested estimated billions of dollars on business intelligence and data warehousing efforts. Most of these initiatives have followed the traditional data- and analytics-centric strategy pushed by the established providers of large software systems. As many enterprises have learned through hard experience, this bottom-up approach has proven to be time consuming and costly, and has rarely delivered either good intelligence or the promised ROI. Now there is a faster, less costly, and more direct pathway to the benefits of business intelligence.

This technology-oriented approach, which is driven by the data-crunching capabilities of the various proprietary software products, requires the construction of the kind of massive data marts that were supposed to "see every detail and answer every question." By assembling a vast warehouse of data, and then applying high-powered analytics to slice and dice that information, the hope was that meaningful performance management insights would eventually percolate to the top of the organization.

All these developments have progressively and greatly increased the reporting and retrieval capabilities of business intelligence systems, although the technology-based support for the task of specifying the information executives want, or need, out of all the available data has not evolved to nearly the same extent. This bottom-up approach has proved to be costly and frustrating, and has not delivered the timely and relevant information key decision makers need. Nor has there been any real improvement in the general understanding of this process, with or without technology to assist.

THE VALUE OF BUSINESS INTELLIGENCE

Every enterprise has a demand for information and this demand is a continuously increasing, constantly changing need for current, accurate, integrated information, often on short or very short notice, to support its activities. The value of business intelligence can be estimated by observing the demand for business intelligence applications, which continues to grow even at a time when demand for most IT products is soft.

Business intelligence involves the coordination of core information with relevant contextual information to detect significant events and illuminate unclear issues. It includes the ability to evaluate business trends, to evolve and adjust to changing situations, and to make intelligent decisions based on uncertain judgments and contradictory information. It relies on exploration and analysis of unrelated information to provide relevant insights, identify trends, and discover opportunities.

Business intelligence certainly has a tremendous potential for helping enterprises become increasingly successful. It has already shown value in many business areas and promises to show increased value throughout a business as it matures. It is the primary support for developing an intelligent learning enterprise.

Business Intelligence Value Chain

The business intelligence value chain is one that supports business goals of an enterprise (see Figure 1.6). A business intelligence value system consists of

- Real-time data warehousing
- Data mining
- Automated anomaly and exception detection
- Proactive alerting with automatic recipient determination
- Seamless follow-through workflow
- Automatic learning and refinement
- Geographic information systems
- Data visualization

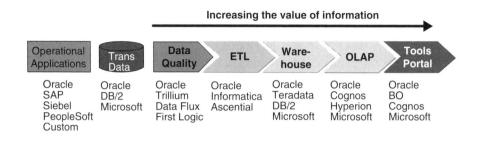

Figure 1.6   Business intelligence value chain

Value is added to the data resource through each step in the business intelligence value chain. The data resource supports the development of information through the information engineering process. Information, in turn, supports the knowledge worker in a knowledge environment, and the knowledge worker supports business intelligence in the intelligent learning enterprise. Business intelligence supports the business strategies, which support the business goals of the enterprise.

Garbage In–Garbage Out

The business intelligence value chain is based on the principle of "Garbage In–Garbage Out." An interesting but obvious aspect of the business intelligence value chain is that any of its levels cannot have better quality than its supporting level. Because the data resource is the foundation of the business intelligence value chain, the quality of any higher level can be no better than the quality of the data resource.

The degree to which business goals are met can be no better than the quality of the data resource used. If the data resource is low quality, it will affect the business intelligence value chain adversely, hence the business goals. But, if the data resource is high quality, it will adequately support the business intelligence value chain, hence the business goals.

Different levels of the business intelligence value chain are the data resource, the information, the knowledge environment, the business intelligence, the business strategies, and the business goals (see Figure 1.7).

The information technology function, responsible for managing the information technology realm including the data resource and the information levels and supporting the human resource realm, must emphasize high-quality data and high-quality information.

The human resources function, responsible for managing the human resource realm including the knowledge environment, and the business

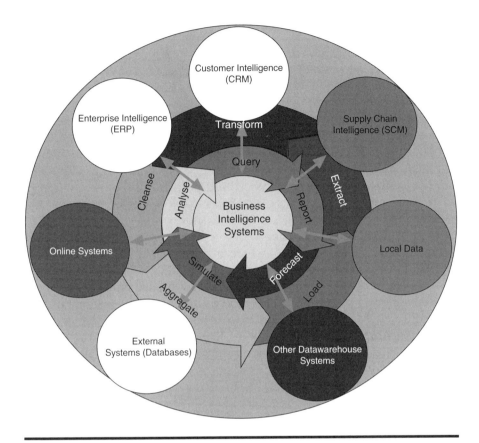

Figure 1.7 BI process circle

intelligence levels, must emphasize knowledge workers and the development of an intelligent learning enterprise. Business managers and executives should be responsible for managing the business realm including business strategies and business goal levels.

The value begins with a high-quality data resource that is the raw material for preparing high-quality information. The value continues with high-quality information supporting knowledge workers and an intelligent learning enterprise. Finally the value continues to the support of business strategies and goals.

Similarly, the business information demand begins with the business goals and strategies, continues down through the business intelligence and knowledge worker levels, and through information to the data resource. When the business information demand is well defined down through the business intelligence value chain, the data resource will be better prepared to support that demand.

Need for Business Intelligence

Under continuing economic pressure, most enterprises are questioning and evaluating all their IT projects and efforts, even the ones that are considered most strategic, such as data warehouses. Across the enterprise, the mantra is to do more with less. However, it is critical that enterprises reevaluate their data warehouse strategies and emphasize the need to reduce redundancy and complexity while also increasing the total value of opportunity. IT organizations continually strive to minimize the number of tools they are responsible for deploying, maintaining, and upgrading. Having multiple tools for each specialized purpose becomes counterproductive, especially as an organization attempts to scale, enhance, and maintain the business intelligence infrastructure.

One step in that direction without compromising on productivity is the move toward convergence, which makes it easier and cost effective for clients to purchase and maintain software assets by minimizing the number of vendors with whom their IT department works. Enterprises can reduce or limit the number of vendors and technologies they use for many different business intelligence activities by opting for more complex solutions incorporating all or most of these activities. Thus the starting point for managing costs is to pick a comprehensive business intelligence environment at the outset.

Why Business Intelligence?

- Business is changing.
- Decision making occurs more frequently.
  - At all levels in the organization.
  - Trend toward "democratization" of information and broadening of decision-making responsibilities.
- No longer a "nice to have" driver is the need to unlock the value within the organization and understand the customer better.
- Most of the data is stored in the legacy system, which is difficult for end users to access.
- The data stores were designed for transaction processing not ad hoc reporting.
- Obtaining the data or a report usually requires waiting for a programmer to either develop the report or provide a customized download program.
- All of the data may not be consistent as of the same point in time.
- There may not be enough copies of the data kept for historical reporting in the operational systems.
- End users do not have the knowledge of what is kept in the existing data stores.

Table 1.6   Business Intelligence Drivers

Make better business decisions
  Increase organizational credibility
Gain timely and accurate insight into business operations and processes
  Real time data integration
Regulatory compliance
  Patriot Act, Bank Secrecy Act/Anti-Money Laundering Act
  Sarbanes–Oxley, Basel II
Optimize operational efficiencies
  Information analysis for LOB and business units
Data as asset
  Integrated Data = Efficient + Intelligent Business

Business Intelligence Drivers

- Make better business decisions—intelligent analytics.
- Multiple growing lines of business and business units.
- Gain timely and accurate insight into business operations and processes.
- Increase organizational credibility and optimize operational efficiencies.
- Data as asset.
  - Need for data consolidation.
- Real-time data integration.
  - Integrated Data = Efficient + Intelligent Business.
- Rapid implementation.
- Minimized total cost of ownership.
- Regulatory compliance.
  - Patriot Act, Bank Secrecy Act/Anti-Money Laundering Act Sarbanes-Oxley, Basel II.

Benefits of Business Intelligence

- Improved business efficiency and productivity.
- Business relationships are enhanced.
- Increased business value is generated.
- Reduction of costs.

Business Intelligence Not Just Data

Today's business managers need factual real-time information at their fingertips to anticipate changes and make proactive decisions. Business intelligence and data management help business managers make better decisions by converting data into useful information. By setting goals and monitoring key performance indicators, successful enterprises instill a performance management culture throughout their organizations, with continuous monitoring and analysis driving corrective actions when necessary. Business intelligence provides organizations with true insight into the critical factors that drive their businesses. Implementation of business intelligence platforms allows organizations to address their true business issues with better, faster, and more reliable information distributed throughout the enterprise.

Business intelligence systems combine data gathering, data storage, and knowledge management with analytical tools to present complex and competitive information to planners and decision makers. Implicit in this definition is the idea that business intelligence systems provide actionable information delivered at the right time, at the right location, and in the right form to assist decision makers. The objective is to improve the timeliness and quality of the input to the decision process, hence facilitating managerial work.

The traditional definition of business intelligence is technology that allows organizations to transform data stored in core business systems into meaningful information. It lets users query and analyze databases to uncover key issues that affect their businesses, ultimately helping people make better, more informed decisions. Functions such as ad hoc reporting, OLAP analysis, and data mining provide this capability from different perspectives; that is, they serve the needs of various audiences, from the nontechnical business user to the power user.

Data Warehousing

Traditionally enterprises wishing to deploy business intelligence applications create a data warehouse. The data warehouse might contain enterprise data from a variety of software applications, business units, geographies, or departments. On a periodic basis, data from various sites around the enterprise would be extracted, transformed, and loaded into the data warehouse. Business users could then receive predefined reports on a regular schedule or could request special ad hoc reports to be created.

The advantage of the data warehouse is that users can query data from across the enterprise. Using online analytical applications (OLAP), analysts create complex multidimensional analyses and deliver meaningful insights to business users that otherwise might not be readily apparent. Although the value of data analysis is indisputable, there are three major disadvantages to the data warehouse approach:

1. The cost and complexity of establishing a data warehouse environment is substantial. Deployment and management costs easily run into millions of dollars, and these environments require well-trained, dedicated staff.

2. Because information is drawn from various repositories across the enterprise, there is a need to extract, transform, and load the data into the data warehouse. This process takes time, so users querying the data warehouse will not receive the freshest, most up-to-date operational data. In addition, requests for additional information not yet in the data warehouse can take weeks or months to complete.
3. The tools used to analyze data residing in a data warehouse are designed for the power user or analyst. They are not well suited for end users, managers, or executives. Thus these individuals must request the services of a developer or database analyst to acquire the reports they need. Again, this causes delays and users do not receive the freshest operational data available.

In the end, enterprises deploying data warehouses trade off data freshness and availability for the ability to perform complex multidimensional analyses on data pulled from across the organization. For most enterprises, this is not an acceptable trade.

Real-Time Reporting

An increasing number of organizations need to report on real-time operational data to make informed decisions and stay competitive. The data warehouse architecture is not appropriate for organizations or managers who need access to up-to-the-minute operational data for real-time decision making. Waiting days or weeks for data to be accessible in the warehouse is not tenable operationally. Most enterprises would gain a competitive advantage if they reduced the time spent collecting and responding to information. Enterprises must pursue a data delivery strategy that delivers information to end users quickly and efficiently.

In these cases, enterprises frequently deploy business intelligence applications against their live production databases. Despite the clear advantages to reporting on real-time operational data, deploying business intelligence applications on production databases carries significant operational costs as well. The major problem with this architecture is that business intelligence applications share database infrastructure with a business application. This aspect of the subject is discussed in the following chapter.

NEW ERA OF ENTERPRISE TECHNOLOGY

Business intelligence is leading enterprises into a new era, an era where the business users have more control over how they analyze, report, and manage the performance of the business. Business users are no longer dependent on IT to support their every need. As a result, they are making faster and better decisions.

IT organizations are now free to do what they do best: manage the technology and core transaction systems of the organization. IT can manage the hardware, security, stability, maintenance, and integration of the technology infrastructure, which is the lifeblood of the organization.

The distinction of roles that BI created is now allowing IT and business to start living together in peace and prosperity. Neither is lacking in ways to add value to their organizations. With these roles and responsibilities being clearer, new opportunities are now presenting themselves, opportunities that were not possible sometime back.

Business Intelligence Solutions

Business intelligence is perhaps among the most commonly used but misunderstood technology terms. Some software vendors use it to describe their generic reporting templates, whereas others choose to define it as a byproduct of infrastructure strategies such as data warehousing (see Table 1.7). For business managers, however, business intelligence is just that: information that helps them in doing business.

Whether they know it or not, business managers are dabbling in business intelligence every day. Most of them need the overnight polling numbers from multiple branch locations, or sit in on weekly sales or production reporting meetings. However, the methods by which those reports are created are typically human-intensive, requiring spreadsheets and presentation programs. Many organizations, hence their managers, use the default reporting and search tools that came with their enterprise applications such as Customer Relationship Management (CRM), Sales Force Automation (SFA), accounting, and other departmental productivity tools. Although these tools provide some level of functionality, they pale in comparison to the high degree of information access and analysis that can come from a business intelligence solution. And because managers have a vested interest in how they access the data, many lines of business managers and executives are taking a more active role in planning and selecting business intelligence solutions to ensure effective management and competitive differentiation in a highly competitive marketplace.

True business intelligence can only be culled from comprehensive information, the information that spans across the range of functional silos that make up the enterprise, organizes for clarity and consistency, provides true decision support, and lays the foundation for competitive superiority and action. Providing that information requires that all operational data be available for analysis, including the vast amounts held in ERP, CRM, and SCM applications. Each business is unique with each solution tailored to the organization's own requirements and resources and fully leveraged

Table 1.7   Enterprise Applications Vendors

Application servers	WebLogic, WebSphere, ColdFusion, Vitria, Active eCommerce Server Transact, SiteServer, net.Commerce, Oracle
CRM	Siebel, Salesforce.Com, Peoplesoft Crm
Data warehouse tools	Informatica, Business Objects, Ab Initio, Brio, Cognos, Microstrategy, Datastage, Sas Enterprise Miner, Spss
Database warehouse data	Oracle 8x, Sybase, DB2, PROGRESS, Informix, SQL Server 2000, Rebbricks, Teradata
Distributed computing	CORBA, RMI, COM, DCOM, EJB
EAI—middleware	TIBCO, SeeBeyond, MQ Series, Crossworlds, Talarian
ERP	SAP, JD Edwards, Peoplesoft, BAAN, Oracle
GUI	VB, VC++, Visual age for Java, HTML, DHTML, Java (AWT/SWING), FORMS
Industry protocols	XML, Java, JMS, EJB, OFX, FIXML, FinXML, FpML, SWIFT, HIPPA, EDI, BPML, BPQL, ebXML, SOAP, BIZTALK, UDDI, XAML, WSDL
Internet technologies	Java, J2EE, JSP, ASP, EJB, Servlets, JavaBeans, XML, JavaScript, Perl, CGI, .NET
Legacy systems	AS400, Mainframe, DEC, VAX, VMS, Tandem
OLAP servers	Essbase, IBM DB2 OLAP Server, SQL OLAP Services, SAS/MDDB, Seagate HOLOS
OLAP tools	Oracle Express Suite, Business Objects, Microstrategy, Cognos Powerplay/Impromptu, SAS, Web Intelligence, MetaCube
Operating systems	UNIX (Sun, AIX, HP), Windows NT/2000, Linux
Programmable languages	JAVA, C++, C, C#, SmallTalk, COBOL
SCM	i2, Oracle
Testing tools	Win Runner, Load Runner, Test Director

earlier IT investments. Organizations must help make an informed buying decision when it comes to selecting business intelligence applications or deciding business intelligence solutions by evaluating advanced capabilities available and their own data access and analysis needs.

Ideal Business Intelligence Applications

Business intelligence applications are decision support tools that enable real-time, interactive access, analysis, and manipulation of mission-critical enterprise information. These applications provide users with valuable insight into key operating information to quickly identify business problems and opportunities. Users are able to access and leverage vast amounts of information to analyze relationships and understand trends that, ultimately, support business decisions. These tools prevent the potential loss of knowledge within the enterprise that results from massive information accumulation that is not readily accessible or in a usable form.

Business intelligence applications must allow businesses to leverage their information assets as a competitive advantage while they allow businesses to better understand the demand side of the business and manage customer relationships. Also they must be capable of allowing enterprises to monitor results of change, both positive and negative.

The desired outcome of business intelligence projects is the continuous improvement of the enterprise through timely information that enhances decision making. Business intelligence applications should enable the enterprise to become proactive and agile by delivering information used to:

- Support internal enterprise users in the assessment, enhancement, and optimization of performance and operation.
- Deliver critical business information to end users about value chain constituencies, such as customers and supply-chain partners.

Fast Decisions Faster

The primary purpose of business intelligence applications is to facilitate better decisions. Better decision making very heavily relies on high-quality up-to-date data delivered to the right managers in a timely and easy-to-understand manner.

Businesses are becoming more focused on the bottom line, with managers more concerned than ever about the cost and benefit of every function, while the pace of business is speeding up as customer demands become more intense and competitors move more quickly than ever to meet their needs.

The combined impact of these factors makes the real-time measurement of the performance of critical business functions, and the proactive communication of these measurements on a continuous basis to managers best positioned to take immediate corrective actions, a necessity. Currently, such systems are generally focused on one operational area, as opposed to overall corporate performance.

Although top management is unlikely to make a snap corporate strategy change based on a real-time performance indicator on an executive dashboard, thousands of lower-level decisions being made continually across most organizations could benefit from real-time performance-based analytics.

Integrated Business Intelligence Solutions

An integrated business intelligence solution gives enterprises better control of their business processes, because they can get instant insight into the events and activities in and around their organizations. That allows business managers to make informed decisions about strategic, tactical, and operational topics. The challenge, however, is that important data such as customer orders, inventory, production levels, and financial accounting reside on multiple servers and across various databases, and that entire infrastructure is generally in the domain of the IT department. The most strategic way to get this information out of the IT department and into the hands of managers who need it is through the greater integration of advanced business intelligence applications. The integration of business intelligence resources can create a single window into all the enterprise applications.

In some enterprises, business managers may be occasionally consulted or informed regarding strategic technology decisions. Most managers, however, report that their interaction with their IT department is generally more tactical and reactive. Truly integrated business intelligence does not require human intervention. In fact, it excels without it. The absence of human intervention in the integrated business intelligence solution also means a greatly reduced reliance on technical support and manual integration of multiple vendor tools to create a piecemeal solution. Integrated business intelligence provides instant access to live information from multiple data sources throughout an organization.

Imagine it: up-to-the-minute reports on what is being purchased, when something is being delivered, or how a region is doing along with the analytical capability to compare numbers across time, locations, or other market segments.

Enterprise Business Intelligence Suite (EBIS)

Integrated business intelligence suites represent an evolutionary step in the delivery of business intelligence solutions from previously disjointed query and report offerings. These suites provide integration of query, reporting, and online analytical processing functions so that users are able to access databases directly and produce reports for display, printing, electronic distribution, or further analysis. A full-scale Enterprise Business Intelligence Suite (EBIS) supports enterprise reporting in which reports are produced for widescale distribution, both within and without the enterprise.

An EBIS must be able to deliver reports to thousands of users who would require little or no training because of the EBIS's ease of use and its ability to support scheduled reporting. The EBIS server should allow for multiplatform deployment and provide tuning and management tools to ensure reliable, redundant, large-scale operation. Another prerequisite for widescale deployment is the ability to deliver a thin-client desktop that provides full-client reporting functionality.

Consumers of information provided by an EBIS have many requirements, but the majority of BI recipients require only basic information, for example, credit-card balances and transactions for consumers or the number and types of widgets produced by a machine on the shop floor of the plant for production control managers. Providing timely information to this majority empowers employees at all levels (not just managers) to make decisions. Therein lies the benefit and payback of an EBIS: improvements in business processes and better management of key functions throughout the enterprise because more employees are receiving timely information on the critical aspects of their jobs.

The essence of enterprise reporting then is the ability of an EBIS to deliver information efficiently and reliably to thousands or tens of thousands of users both within and beyond the enterprise. Information delivered to this majority usually does not contain more than 10 percent of BI content (information that needs further analysis). Knowledge workers such as HR managers or production planners may require up to 30 percent of the information from their standard or predefined reports that need to be further analyzed.

The EBIS provides the ability to enter parameters at runtime, enables users to tailor their reports, drill down on specific data, or retrieve myriad combinations of data from one report screen. This data, which can be repeatedly or recursively accessed in different ways, can then be further analyzed using an integrated OLAP engine and can provide a very high degree of analytical power that will satisfy the needs of all but the most demanding business analyst.

An EBIS must also provide ad hoc reporting to thousands of potential users to enable them to query their data directly by developing simple reports on the fly. This function can be provided by parameterized reporting, which uses a predeveloped screen format with pull-downs. These can be used in a number of screening, retrieval, and command combinations to produce hundreds of different self-service, ad hoc reports from a single screen. The ability to do this requires little or no training and thus drives down the costs of information delivery to employees, partners, suppliers, and customers accessing the suite.

Although function is paramount for providing easy access to corporate information for the broadest possible base of users, there must be a scalable Web server driving the application. Oddly enough, about half of the reputed EBISs on the market today were not designed originally for the Web, but were spawned as client/server tools. The significance of integrated suites is that they represent the convergence of business intelligence and enterprise reporting. Enterprise reporting is the high-volume output of standard reports to a large audience, but which lacks the ad hoc and OLAP facilities of a BI tool. The more prominent BI tools have client/server architectures and are constrained when it comes to high-volume enterprise reporting and information delivery. High-volume reporting may require advanced report bursting, selection, and archiving that simply do not exist among BI products that were originated for the client/server environment. A Web-designed integrated architecture provides for a high degree of scalability and manageability whereby administrators are aided in the deployment of Web functionality without adding new resources.

Summing up, the types of query facilities required for an integrated suite are

- Ad hoc
- Scheduled
- Production
- OLAP
- Parameterized self-service

The other design requisites for a full-function EBIS are

- Scalability
- Access to multiple databases
- Desktop, thin client, and portal integration

BUSINESS INTELLIGENCE APPLICATIONS

Recent significant changes in the business environment, for example, the globalization of markets, ever-faster technological developments, and the increased importance of knowledge-based assets, have brought about new managerial challenges. BI is the tool that organizations can leverage to gain competitive advantage in this dynamic era of the performance-oriented modern business environment.

Technology providers and vendors have devolved many applications and tools using exceptional features and flexibility offered by BI to solve different problems from different decision-making points of views. Simply stated, today business intelligence can be applied in almost all industries and sectors. BI applications provide effective and efficient monitoring and control of operative processes while they assist in decision making and guide towards strategic direction. BI applications can be effective across most processes, functions, and departments of an organization, from order processing to purchasing up to financial accounting and resource management. Some of such applications are shown in Figure 1.8.

GOVERNMENT	FINANCIAL SERVICES	RETAIL	TELCO
Service to the Citizen	Compliance Reporting	Store Operation Analysis	Fraud Analysis
Homeland Security	Portfolio Analysis	Customer Loyalty Programs	Churn Analysis
E-Government	Customer Statements		Improving Response Times
	Customer Profitability	Collaborative Planning and Forecasting	
Enforcement & Regulation	Wire Transfer Alerts		Traffic Analysis
	Branch Office Scorecards	Loss Prevention	Product Affinity/ Bundling
Human Capital Management	Customer Acquisition, Retention, Profitability	Supply Chain Optimization	
Information Dissemination			

Figure 1.8   BI applications

Distribution and Customer Data Analysis

Increasing globalization is intensifying competition at such a rate that an extensive individual analysis of customer relationships and the product mix is essential. Visualization of distribution data creates transparency. Correlations between marketing trends and distribution results will be visible and will give an enterprise a range of options. An enterprise can also get in-depth understanding about its customers and can identify profitable customer relationships and extend them continuously.

Cost Accounting and Controlling

The basic requirement for the operational management of an enterprise is cost transparency. An enterprise can identify cost drivers and can study historical and emerging patterns by tracking the development of costs that can be used for cost minimization and optimization of the value chain.

Breakeven Analysis

The breakeven analysis offers an important tool for planning and controlling an organization. The organization can analyze fixed and variable costs as well as the profitability of clients and products. Thus, it can create a range of products more profitably and make profound make-or-buy decisions.

Order and Production Data Analysis

The record and the specific analysis of the production process with the help of key data permit detailed capacity planning. An organization can identify disturbance variables at an early stage and be able to compensate for them in terms of an active quality management.

Supply Chain Management Monitoring

Knowledge about correlations of the entire supply chain holds enormous potential. First, the organization can analyze customer requirements in detail. With the help of diverse key data (supplier assessment) transparency about the behavior of suppliers can be created. With the analysis of stock data the organization can optimize supply chain, turnover rate, purchase order quantity, and delivery times, and thus reduce costs and delivery times.

Balanced Scorecard

The balanced scorecard is a management tool for the realization of strategy controlling. By means of KPIs of the scorecard, management can monitor the development of the organization permanently. This early-warning system gives signals needed to control an organization in a fast and effective way.

Process Performance Measurement Systems

Today, milestone-oriented process flows are established in many divisions of an organization. In the past these work flows were increasingly optimized, so that a higher degree of effectiveness and efficiency could be reached. However, in many cases a centralized process control is still missing. In this case, the use of business intelligence produces relief by definition of key performance indicators and a consistent appraisal and presentation of these KPIs by a central MIS system.

These are only some of the applications of BI. In the real world there can be numerous areas where BI can play a vital role in improving or enhancing performance. BI application allows organizations to make the right decisions on the basis of safe information in practically every organizational department.

2

BUSINESS INTELLIGENCE: ESSENTIALS

To make the kind of decisions that keep business competitive, your enterprises must be able to sort through an increasing volume of information quickly while maintaining high performance levels.

A critical component for the success of the modern enterprise is its ability to take advantage of all available information. In the past few years there has been an explosion of information, internal as well as external, available to an organization and the trend seems to continue or may even strengthen in days to come. Data environments are becoming more and more complex as the data is useful only if organizations have the systems and tools to harness it and turn it into information for evaluation and decision-making purposes. With the constant increase in the volume of data available and the amount of information they manage, organizations are finding this challenge of making data work in their favor more difficult.

Information is the lifeblood of organizations today, and the faster it flows, the healthier a business will be. Corporate culture and the increasingly dynamic nature of the global marketplace have also placed a growing importance not only on the quantity and quality of information an organization gathers, but also on the speed at which it can be obtained and shared. As a result information accessibility and delivery has shifted from point in time to real-time. At the same time, a technology-savvy public is requesting more information more frequently. This leads to the compelling need for integrating the various islands of information that result from disconnected applications and data sources, and making them available for analysis, monitoring, and reporting across the enterprise.

FACTS, DATA, INFORMATION, AND KNOWLEDGE

Data, information, and knowledge are three evolutionary stages of a common resource, "facts," but vary widely based on their scope and value to the enterprise. Facts are numbers, characters, character strings, text, images, voice, video, and so on. Facts are just representations of actual or hypothetical events, parameters, and instances.

Data is the individual raw facts that are out of context, have no meaning, and are difficult to understand. Data in context is facts that have meaning and can be readily understood. It is the raw facts with meaning and understanding, but not yet information.

Information is a set of data in context that is relevant to one or more entities at a point in time or for a period of time. Information is data in context with understanding about what the facts mean. Information is data characterized by meaning, relevance, and purpose. A set of data in context is a message that only becomes information when someone readily accepts that message and it is relevant to his needs. Information must have relevance and a timeframe.

Knowledge is cognizance, the fact or condition of knowing something with familiarity gained through experience or association. It is the acquaintance with or the understanding of something, the fact or condition of being aware of something, and of apprehending truth or fact. At present, context knowledge can be tacit, explicit, or organizational in nature.

- *Tacit knowledge* is the knowledge that is in people's heads or the heads of a community of people, such as an enterprise. It is what makes people smart and act intelligently.
- *Explicit knowledge* is knowledge that has been rendered explicitly to a community of people, such as an enterprise, and is what they deem to know.
- *Organizational knowledge* is information that is of significance to the enterprise, is combined with experience and understanding, and is retained. It is information in context with respect to understanding what is relevant and significant to business issues.

BI ENABLING ENVIRONMENT

Business Intelligence (BI) grew out of the need to pull information together from multiple disparate systems as well as to retain a historic perspective of how business operations change over time. Many organizations believe that decisions and operational adjustments made today should be influenced by observing trends and patterns from the past. Making the best actionable decisions requires the ability to draw upon information and solutions that make up one's business intelligence framework.

Despite efforts to improve effectiveness, many organizations realize their existing application software (ERP, CRM, POS, etc.) does not always live up to the promises associated with better business insight. Business intelligence, unlike enterprise applications such as CRM and ERP, does not require the "massive overhauling" of systems or processes, nor does it require organizations to "rip and replace" existing infrastructure. Business intelligence can fit very nicely into existing systems and can be implemented both on a divisional level as well as across the organization.

Technology Landscape

Many factors have influenced the quick evolution of the BI discipline. The most significant set of factors has been the enormous forward movement in the hardware and software technologies. Sharply decreasing prices and the increasing power of computer hardware, coupled with advanced functionality and ease of use of today's software, has made possible quick analysis of hundreds of gigabytes of information and business knowledge.

The use of technology by managers and top executives has increased significantly. They have decisively moved beyond using the personal computer for e-mail. This hands-on use of information and technology by upper management has facilitated the sponsorship of larger projects such as data warehousing.

Alongside the availability of key enabling technologies, these fundamental changes in the nature of business over the past decade have played a central role in the evolution of BI and DW technologies. Some might even argue that these changes in business have led the technology to its current state.

Hardware Prices Go South

The most important factor in the evolution of BI and DW technologies has been the sharply increasing power of computer hardware. Along with the increase in this power, prices have fallen just as sharply.

Gordon Moore, co-founder of Intel, predicted that the capacity of a microprocessor would double every 18 months. This has not only held true for the processor, but also for other components of the computer. Although desktop computers today are more powerful than the mainframes of yesterday, an inexpensive server possesses power that was difficult to imagine just a decade ago.

Sophisticated processor hardware architectures such as symmetric multiprocessing have come to mainstream computing with inexpensive machines. Higher-capacity memory chips and storage devices along with advancement in storage and retrieval systems are now available at very low prices. Computer buses such as PCI and controller interfaces such as

Ultra SCSI have made I/O incredibly fast. Just two decades ago, it would have taken a roomful of disk drives to store information that can now be easily stored on a single one-inch-high disk drive.

Desktop Power Increasing

The personal computer has become a hotbed of innovation during the past decade. The personal computer was initially used for word processing and other minor tasks with no links to primary analytical functions. With the help of innovations such as powerful personal productivity software, easy-to-use graphical interfaces, and responsive business applications, the personal computer has become the focal point of all computing today.

Powerful desktop hardware and software have allowed for the development of the client/server or multi-tier computing architecture. Almost all data warehouses are accessed by personal computer-based tools. These tools vary from very simple query capabilities available with most productivity packages to incredibly powerful graphical multidimensional analysis tools. Without the wide array of choices available for data warehouse access, data warehousing would not have evolved so quickly.

Increasing Power of Server Software

Server operating systems such as Windows NT and UNIX have brought mission-critical stability and powerful features to the distributed computing environment. The operating system software has become very feature-rich and powerful as the cost has been going down steadily.

With this combination, sophisticated operating system concepts such as virtual memory, multitasking, and symmetric multiprocessing are now available on inexpensive operating platforms. Operating systems such as Windows NT have made these powerful systems very easy to set up and operate reducing the total cost of ownership of these powerful servers.

Intranets and Web-Based Applications

The most important development in computing since the advent of the personal computer is the explosion of Internet and Web-based applications. Somewhat after the fact, the business community has quickly jumped onto the Internet bandwagon.

Another exciting field in the computing industry today is the development of intranet applications. Intranets are private business networks that are based on Internet standards, although they are designed to be used internally. The Internet/intranet trend has very important implications for data warehousing applications.

- First, BI applications can be made available worldwide on a public or private network at a much lower cost. This availability minimizes the need to replicate data across diverse geographical locations.
- Second, this standard has allowed the Web server to provide a middle tier where all the heavy-duty analysis takes place before it is presented to the Web-browsing client to use.

BUSINESS LANDSCAPE

Another very significant influence on the evolution of BI technology is the fundamental changes in the business organization and structure during late 1980s and early 1990s. The emergence of a vibrant global economy has profoundly changed the information demands made by corporations worldwide. Enterprises have found markets for their products globally while competing with other enterprises in vastly different cultures and economic environments. The mergers and acquisitions of businesses have crossed the country boundaries.

Economic Factors Leading to Consolidation

The economic downturn of the late 1980s led many global enterprises through a remarkable period of consolidation. Phenomena such as "business process reengineering" and "downsizing" forced businesses to reevaluate their business practices. Many industries went through prolonged periods of consolidation and reinvention. During this period, simple economics forced the businesses to identify their core competency areas and shed businesses that were not profitable. These economic factors have played an important role in the evolution of performance improvement systems, including BI and DW.

Emergence of Global Enterprises

Competition from emerging economies has forced large enterprises to become lean and efficient. The emergence of this global economy has led to the migration of manufacturing to less expensive and less restrictive countries. Along with these opportunities they present a very volatile business climate and economies that are nearly impossible to predict.

Businesses have not only focused on building products worldwide, but they have also changed their organizations to sell products around the globe. This globalization of business has increased the need not merely for more continuous analysis, but also to manage data in a centralized location. The process of rolling up manufacturing and sales data from far-flung business units has now started to affect a much larger number of corporations. Businesses now need to continuously make "build or buy" decisions.

Globalization of business has made the consolidation of data in a central data warehouse more complicated. Factors such as currency fluctuations and product customization for different markets have added complexity to data warehousing, making the analysis much more complicated.

Emergence of Standard Business Applications

Another factor that is fast becoming an important variable in BI equations is the emergence of vendors with popular business application suites. Led by software vendors such as Oracle and SAP AG, flexible business software suites adapted to the particulars of a business have become a very popular way to move to a sophisticated multi-tier architecture. Other vendors have likewise come out with suites of software that provide different strengths, but have comparable functionality.

End Users More Technology Savvy

One of the most important results of the massive investment in technology and movement toward the powerful personal computer has been the evolution of a technology-savvy business analyst. Even though the technology-savvy end users are not always beneficial to all projects, this trend certainly has produced a crop of technology-leading business analysts that are becoming essential to today's business. These technology-savvy end users have frequently played an important role in the development and deployment of data warehouses.

They have become the core users that are first to demonstrate the initial benefits of BI and DW. These end users are also critical to the development of the data warehouse model: as they become experts with the data warehousing system, they play a very important role of mentoring other users.

Word processing and spreadsheets were the first applications to be effectively used on personal computers. In fact, the spreadsheet is said to be the killer application that led to widespread deployment of personal computers. The charting functions from a spreadsheet represent one of the most extensively used business analysis and presentation functions. The new pivot tables available in popular spreadsheets have allowed for simple multidimensional analysis. The aggressive use of inexpensive personal productivity software has led to the use of more robust reporting and analysis tools along with more powerful desktop database engines. These powerful tools are now targeted more toward the end user and often require very little training for simple applications.

Management More Information Conscious

Many factors affect the heightened awareness of trends in information technology among middle and upper management levels. Unlike a decade ago, information technology now is nearly universally accepted as a key strategic business asset. Many mid- and upper-level managers that have risen through the ranks over the past decade have invariably made their mark with successful technology investments. As a result, they tend not to shy away from risking resources on new and emerging technologies. The explosive use of the Internet has greatly aided in the managers' awareness of technology trends. The Internet is now being used to conduct business transactions, but its greatest asset to this date has been dissemination of information. Today, executives can not only review various sources of industry trends, but also can readily find case studies and vendor information.

CHANGING DATA LANDSCAPE

Data, complex and cumbersome as it may seem on the surface, is one of the most strategic assets an organization possesses in the present knowledge economy. Organizations of all sizes and sectors can shorten business cycles, increase responsiveness to changing business conditions, and gain a competitive edge by finding faster and more effective ways to turn data into relevant actionable information.

Faced with increasing volumes of data, today's enterprises need to identify, classify, access, manage, integrate, and analyze their vital business information in order to measure their performance (see Table 2.1). However, most organizations hold their vital information in a variety of incompatible data stores spread across the enterprise. In some organizations critical customer information can be held in up to 40 different data "silos."

Over the past decade, managers have seen an exponential increase in data that they are expected to process: demand forecast information from Sales Force Automation (SFA) systems, market information published by exchanges, the number and complexity of financial instruments available to hedge, logistical data from Supply Chain Management (SCM) systems, and so on. And there's no end in sight: technologies such as Radio Frequency Identification Technology (RFID) will drive an even faster pace of data growth.

Data System History

In the 1970s virtually all business system development was done on IBM mainframe computers using tools such as COBOL, CICS, IMS, DB2, and the like. The 1980s brought in the new minicomputer platforms such as

Table 2.1 Journey from Data Management to Data Analysis

1960s
Data collection, database creation, IMS, and network DBMS

1970s
Relational data model, relational DBMS implementation

1980s
RDBMS, advanced data models (extended-relational, deductive)
  and application-oriented DBMS (spatial, scientific, engineering)

1990s
Data mining and data warehousing, multimedia databases, and Web
  technology

2000s
Business intelligence, enterprise business intelligence suites with
  a wide range of applications including but not limited to production
  reporting, portal connections, ad hoc reporting, and OLAP viewing

AS/400 and VAX/VMS. The late 1980s and early 1990s made UNIX a popular server platform with the introduction of client/server architecture.

Despite all the changes in the platforms, architectures, tools, and technologies, a remarkably large number of business applications continue to run in the mainframe environment of the 1970s. One of the key reasons behind this is that over the years these systems have grown to capture the business knowledge and rules that are incredibly difficult to carry to a new platform or application.

These systems, generically called legacy systems, continue to be the largest source of data for analysis systems. The data that is stored in DB2, IMS, VSAM, and so on for the transaction systems ends up in large tape libraries in remote data centers. An institution will generate countless reports and extracts over the years, each designed to extract requisite information from the legacy systems. In most instances, IS/IT groups assume responsibility for designing and developing programs for these reports and extracts. The time required to generate and deploy these programs frequently turns out to be longer than the end users think they can afford.

Prior to the start of the information age in the late 20th century, businesses sometimes took the trouble to struggle to collect data from non-automated sources. Businesses then lacked the computing resources to properly analyze the data, and often made commercial decisions primarily on the basis of intuition. The advent of the Internet has increased the complexity along with the pace of today's business environment, creating a tremendous surge in the number of business transactions and the amount of business data.

As businesses started automating more and more systems, more and more data became available. Applications that automate key operations such as customer relationship management, supply-chain management, and product life-cycle management still continue to proliferate while the amount of data organizations must manage grows exponentially.

However, collection remained a challenge due to a lack of infrastructure for data exchange or to incompatibilities between systems. Reports on the data gathered sometimes took months to generate. Such reports allowed informed long-term strategic decision making. However, short-term tactical decision making continued to rely on intuition.

In modern businesses, increasing standards, automation, and technologies have led to vast amounts of data becoming available, leading to a data explosion problem (see Table 2.2). Automated data collection tools and mature database technologies have led to the creation of tremendous amounts of data. Today enterprises are literally drowning in data, but starving for knowledge. This data is stored in databases, data warehouses, and other information repositories.

Improved ETL (Extraction, Transformation, and Loading) and even recently enterprise information integration tools have increased the speedy collecting of data. OLAP reporting technologies have allowed faster generation of new reports that analyze the data. Enterprises are becoming increasingly "knowledge centric" and have been investing in technology in an effort to manage the information glut and to glean knowledge that can be leveraged for a competitive edge.

Most of the strategic initiatives such as regulatory compliance, single view of the customer, business performance management, and supply chain optimization require integrated, accurate, reliable, and, most importantly, timely data. Business intelligence is no different, and requires high-quality data as input to deliver business goals as output. Business intelligence

Table 2.2 Key Trends Shaping the Data Landscape

Data volume is doubling every two years and is expected
  to grow at even higher rates in the future
Web browser is becoming the standard interface for business
  intelligence applications
Enterprises are increasingly opening data warehouse
  and business intelligence applications to customers,
  suppliers, and partners
Cost to deploy numerous analytical tools is becoming
  prohibitive for enterprises
Customer-centric analysis will require increasingly robust data
  warehouses and business intelligence architecture

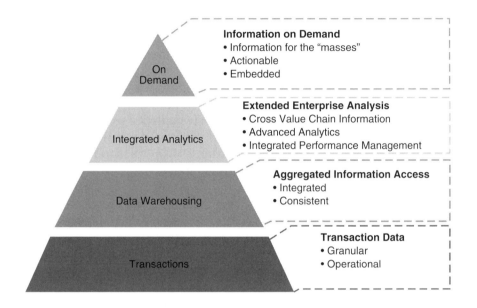

Figure 2.1   Business intelligence framework—access

has now become the art of sieving through large amounts of data, extracting information, and turning that information into actionable knowledge (see Figure 2.1).

DATA QUALITY

The business intelligence value chain is a chain where value is added from the data resource, through each step in the value chain, to the support of business goals. The data resource supports the development of information through the information engineering process. Information, in turn, supports the knowledge worker in a knowledge environment, and the knowledge worker supports business intelligence in the intelligent learning organization. Business intelligence supports the business strategies, which support the business goals of the organization.

Business intelligence requires high-quality information which is derived only from a high-quality data resource. Information quality is a measure of how well the business information demand is met. It is the ability to get the right data to the right people, in the right place, at the right time, in the right form, at the right cost, so they can make the right decisions, and take the right actions. Information quality heavily depends upon data resource quality, which is a measure of how well the data resource supports the current and future business information demand.

Changes in regulatory compliance and the growth of "info-centric" architectures may have recently made data quality issues a concern, but in the past businesses often preferred to leave the job of sorting out occasional inaccuracies of their data to the individual discretion of staff. Organizations must understand the need for, and the value of, a high-quality data resource.

BI users have always been aware of just how dependent they are on the quality of the data they interrogate. However, the low visibility of the data quality issue meant they were often left to act on their own initiatives and the workarounds they developed would often detract from the value of the investment in BI and analytic tools. Technology will support business intelligence, but the real issue is how to achieve and maintain a high-quality data resource to support business intelligence. Any level in the business intelligence value chain has no better quality than its supporting level. Because the data resource is the foundation of the business intelligence value chain, the quality of any higher level can be no better than the quality of the data resource. The degree to which business goals are met can be no better than the quality of the data resource.

The data resource is like the foundation of a house. If the foundation is not level or square, the carpenter fights the error clear to the last shingle on the roof. If the foundation is level and square, the house remains level and square. If the data resource is low quality, it will affect the business intelligence value chain clear to the business goals. But, if the data resource is high quality, it will adequately support the business intelligence value chain.

Bad Data

Business intelligence tools are only as good as the data they use. Bad data already damages business. It can interfere with the processes that take place within a business and the processes that take place between businesses and customers. It can waste marketing resources, damage an organization's reputation, and make it vulnerable to both litigation and fraud.

Data is the most essential working material for business analysis and bad data is a real danger here: an organization that gambles on a new commercial strategy that is underpinned with misleading intelligence can do itself irreparable damage. The simple fact is that poor data results in bad decision making, which can in turn imperil an organization's very strategy. A study done by The Data Warehousing Institute estimated that enterprises lose more than $611 billion every year thanks to bad data.

Data can go bad in many different ways. Irrespective of the individual steps, data traveling across applications and databases introduces not only value but risk. The number of times a data element is moved or touched, either by a person or a system, is inversely proportional to its quality.

Data can also be inaccurate, incomplete, or out of date. It can be used out of context or changed deliberately or accidentally. To maintain data reliability, techniques, methods, and processes that continuously monitor data quality and take corrective measures to improve data consistency are required.

Furthermore, as business needs change, data quality too cannot be static because data itself—and the way people use it—change constantly. Measuring data quality improvements is, therefore, critical. And understanding the evolving needs of the data moving across the data supply chain is paramount.

BI and Data Quality

In the last couple of years the organizations that traditionally took the initiative in data warehousing and BI analytics have started to recognize that the data within their systems is a strategic resource and a valuable business asset. BI is particularly sensitive to data quality hence for data, quality is the key here.

Much of the strategic attraction and value of a BI tool comes from the way it can bring visibility to interdepartmental processes without incurring the considerable headache and expense of investing in operational networks. Enterprisewide visibility is a worthy and valuable goal, but although the concept is an easy one to sell, putting it in place on the ground is a very different matter.

Many of these initiatives started with the need to demonstrate compliance with legislation such as Sarbanes–Oxley or Basel II, yet the need to demonstrate compliance has provided both the incentive and momentum needed to drive the data quality and data integration agenda past the barriers that traditionally held it back. Some enterprises have demonstrated that compliance, which led to considerable improvement in data quality, has resulted in a marked improvement of enterprise performance.

The rise of complex services that use data from multiple strategic partners has similarly driven interest in data quality. Customer-facing service providers are acutely aware of the value of being able to independently assess the accuracy of the data before passing it on to their customers. Similarly, success within a merger or acquisition activity is often dependent on the speed with which each party can understand and exploit the value of the new data assets.

With BI tools becoming more affordable and feature rich than ever before, it is BI that may well deliver the broadest and widest gains from a systematic approach to understanding data quality. Today, BI tools are more effective than ever before. They are intuitively designed, flexible, fast, and invariably packed with rich feature sets. BI has obvious potential as a strategic application, but this potential can only be realized if the organizations know their data assets and how they can be used. Used in conjunction with data quality tools, BI has the potential to change the shape of a market. Used without, at best it is an expensive vanity, at worst, a misleading liability that can harm your business.

BUSINESS INTELLIGENCE FRAMEWORK

Organizations are building business intelligence systems that support business analysis and decision making to help them better understand their operations and compete in the marketplace. Business intelligence consists of sourcing the data, that is, extracting electronic information from text documents, databases, images, media files, and Web pages, using data mining, text understanding, and image analysis techniques to synthesize useful knowledge from collected data, filter out unimportant information, analyze the data, assess the situation, link the useful facts and inferences, identify reasonable decisions or courses of action based on the expectation of risk and reward, develop methodology to identify good decisions and strategies, and then support the decisions made (see Figure 2.2).

BI Building Blocks

Business intelligence is a natural outgrowth of a series of previous systems designed to support decision making (see Figure 2.3). The emergence of the data warehouse as a repository, the advances in data cleansing that led to a single data (or truth), the greater capabilities of hardware and

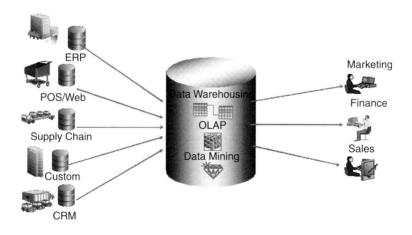

Figure 2.2   BI framework

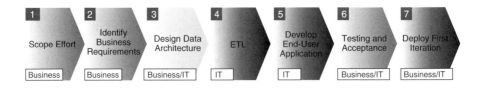

Figure 2.3   Designing a BI system

software, and the boom of Internet and Web technologies that provided the prevalent user interface all combine to create a richer business intelligence environment than was available previously.

Business Intelligence Pyramid

Building a business intelligence system requires intelligent building blocks (see Figure 2.4). A typical business intelligence system can be seen as an informational or technical "pyramid" containing the following elements (also see Figure 2.5).

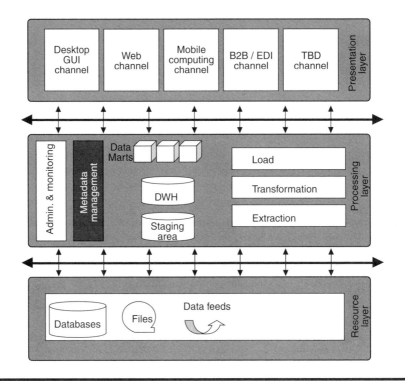

Figure 2.4   BI block framework

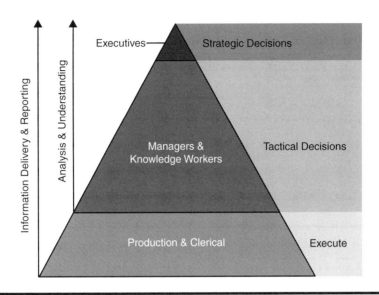

Figure 2.5   Business intelligence pyramid

- Data sourcing systems (legacy systems, Enterprise Resource Planning [ERP] solutions)
- Data enhancement and storage systems (data warehouse/marts)
- Analytic systems (data mining and knowledge discovery)
- Reporting systems (enterprise information system, management dashboard systems, key performance indicator portal, and enterprise portal)

As you work your way up the pyramid, the level of information correspondingly becomes more summarized and the views of information become more strategic.

Business intelligence represents a broad category of applications and technologies for providing access to data to help enterprise users make better business decisions (see Figure 2.6).

Data Sourcing Systems

Data sourcing systems are tools for extracting information from multiple sources of data. The data might be:

- Text documents, for example, memos, reports or e-mail messages
- Photographs and images
- Multimedia or sound
- Tables and lists
- Web pages

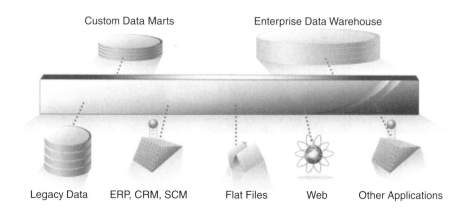

Custom Data Marts   Enterprise Data Warehouse

Legacy Data   ERP, CRM, SCM   Flat Files   Web   Other Applications

Figure 2.6   Data integration

The key to data sourcing is to obtain the information in electronic form. Typical sources of data might include scanners, digital cameras, database queries, Web information, computer files, and automation data to name a few.

Data Integration Framework

The volume of data that IT organizations are managing, the variety of data formats, and users' velocity requirements for access to information have all increased dramatically. Likewise, data integration technology has advanced to meet the more challenging data consistency management requirements created by these demands for access to information.

Project teams continue to employ different tools for the task, centering around extraction, transformation, and loading tools and application integration suites. These two classes of technology continue to converge, leading to contention between proponents of each. Yet the sheer number of choices, usually provided by different vendors, actually compounds the challenges. Business user requirements for access to and manipulation of corporate information dictate requirements for data integration technologies. These requirements fall into five different frameworks.

Replication

Particularly in larger enterprises users are dispersed across many different geographies and time zones. To accommodate all users' needs for access, data for one application may be artificially distributed. In these scenarios, typically, data from the source system is moved on a regular schedule to the receiving target system. Once updates to the source are complete and maintenance is finished, a new replica copy can be delivered to the receiving system again, overlying the previous version. Duplicate copies are created so that one can be maintained offline while users continue to access the alternative copy. In addition, for disaster recovery reasons, duplicate copies of data may be kept in separate locations.

Aggregation

Different users require access to data for their own unique analyses in support of their decision making. Even the same subject area of information may need to be organized and refined in various different ways to meet different analytical requirements. To meet a broad and diverse set of business intelligence requirements, operational data from various sources is consolidated, reorganized, and refined into data warehouses, data marts, operational data stores (ODSs), reporting systems, and other applications to support analytical needs. Here again, data typically moves in a unidirectional manner, from multiple sources into one target. A key characteristic of this pattern is that the data is being transformed or aggregated along the way to meet the specific needs of BI.

Multi-Step Point-to-Point Interfaces

Organizations have a number of siloed (or stovepipe) applications; many of them have been integrated in the past by shipping files from one to another, often by manually writing scripts moving data from one point to another, reflecting that the output of one silo is in fact the input required by another silo. Consequently, many enterprises today have found their IT operations hampered by the resulting spaghetti network of cross-application interdependencies manifested as point-to-point interfaces. However, what these interface networks accomplish is the automation of a multi-step business process and potentially even multiple business processes. Traditionally, this information moved once a day as a file to the receiving system, although in recent years, information is moved as messages throughout the day.

Information Logistics Agent (ILA)

With distributed copies of similar or related data and requirements for reducing the latency of the shared information, IT may have to synchronize various copies of data in a near real-time fashion. This occurs when the processes of the various systems are unrelated, yet there are data elements. Logic must be written to reconcile both formatting (syntax) and semantic variations in the copies. This logic is encapsulated as an Information Logistics Agent (ILA), running independently of any singular application. The ILA essentially represents a new process that is laid over existing data structures to resolve inconsistencies in the data resulting from prior distribution. The risk in this design, however, is that this new logic lives outside any business process or application and creates an alternative method by which the data state is changed.

Direct Real-Time

Senior-level executives need certain pieces of information about business operations available at their fingertips. Usually, the need is for access only (i.e., reading), not updating. Thus, IT must provide online direct access to information sources in such a way that is secure and that will not have too much of a performance impact on the native application and users utilizing the data source (see Figure 2.7).

Data Enhancement and Storage Systems

The data enhancement and storage system is a tool for extracting information from multiple sources, the translation of disparate data into a common, rational data model, and the loading of data into the analytic applications. The data structures are optimal for extracting, transforming, and loading large volumes of data values from the legacy and ERP systems. The resulting data is ready to be analyzed by power users within the organizations or further manipulated within the analytic application layer.

Analytic Systems

The analytic system is a tool for analyzing data. It is the component responsible for BI capable of converting data into information. Some of the tasks made possible by analytic systems are knowledge discovery, situation awareness, and risk assessment.

Business intelligence is about synthesizing useful knowledge from collections of data. It is about estimating current trends, integrating and summarizing disparate information, validating models of understanding, and predicting missing information or future trends. This process of data analysis is also called data mining or knowledge discovery.

The user needs the key items of information relevant to her needs, and summaries that are syntheses of all the relevant data (market forces, government policy, etc.). Business intelligence fulfills user demand by filtering out irrelevant information, and setting the remaining information in the context of the business and its environment providing critical situation awareness. Situation awareness is the grasp of the context in

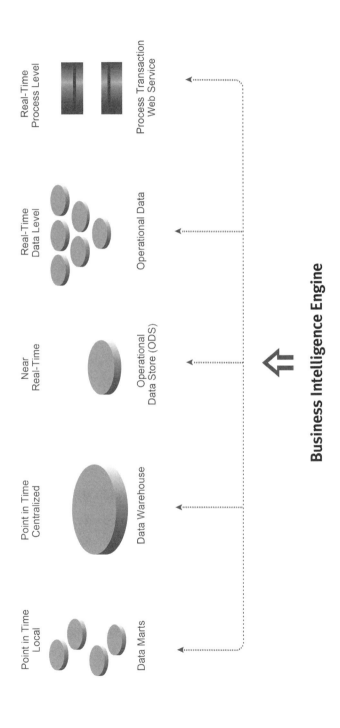

Figure 2.7 BI engine

which to understand available information and make decisions. Algorithms for situation assessment provide such syntheses automatically.

Business intelligence is about empowering organizations to weigh the current and future risk, cost, or benefit of taking one action over another, or making one decision versus another. Business intelligence assists organizations in risk assessment.

Business intelligence is about using information wisely, discovering what plausible actions might be taken, or decisions made, at different times. It aims to keep organizations aware and prepared for important events, such as takeovers, market changes, and poor staff performance, so that preventive steps can be taken. BI seeks to help organizations analyze and make better business decisions and to improve sales or customer satisfaction or staff morale. It presents the information you need, when you need it. It is about inferring and summarizing your best options or choices. Typical analysis tools might use:

■ Probability theory, for example, classification, clustering, and Bayesian networks
■ Statistical methods, for example, regression
■ Operations research, for example, queueing and scheduling
■ Artificial intelligence, for example, neural networks and fuzzy logic

Reporting Systems

Reporting systems allow access to BI analytics with minimal effort. Access can be through widely available browsers, portal solutions, or reports. The system provides the presentation view and interaction with the business analytic applications in a user-friendly manner.

BUSINESS INTELLIGENCE PLATFORM

BI often uses Key Performance Indicators (KPIs) to assess the present state of business and to prescribe a course of action. A BI platform ties into a corporate database platform, has a user reporting and querying interface, and breaks down into five essential layers (see Figure 2.8).

Access Layer

The access layer (Figure 2.9) of the business intelligence framework defines the functions and services to access BI analytics with minimal effort. Access through thin client, widely available browsers, and portal solutions that add value through filtering and organizing capabilities are presented as key strategies within this section of the framework.

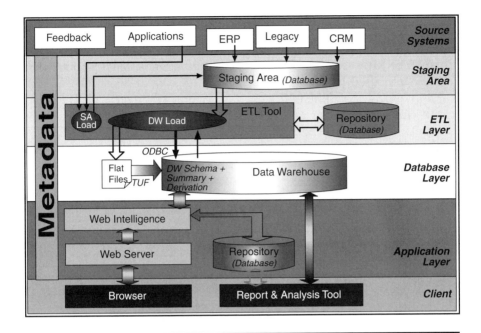

Figure 2.8   BI framework

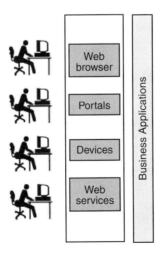

Figure 2.9   Business intelligence framework: access layer

The access layer provides the presentation view and interaction with the business analytic applications. Examples of access layer tools include:

- Web browser—Brio, Business Objects, Cognos, MicroStrategy, SAS Web-based front ends
- Portals—WebSphere Portal, WebSphere Commerce with portlets from IBI, Crystal, Actuate
- Devices—PC, PDA, mobile phone, kiosks, ATMs
- Web services—WebSphere Business Integration

Data Layer

The data layer (Figure 2.10) contains the business intelligence data stores. These data stores should be viewed as single repositories even though they may exist as a set of data stores. Examples of tools to support the data repository layer include:

- Databases—DB2 ESE, DB2 OLAP server, Hyperion Essbase, Oracle, SQL Server, Informix, Red Brick
- Metadata—Data Warehouse Manager, CA Repository, MetaStage

Analytics Layer

The analytics layer (Figure 2.11) of the business intelligence framework defines the functions and services to present solutions to business questions

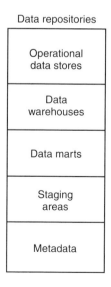

Data repositories

Operational
data stores

Data
warehouses

Data marts

Staging
areas

Metadata

Figure 2.10   Business intelligence framework: data repositories layer

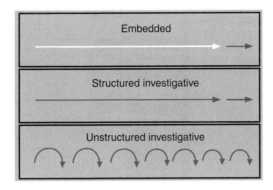

Figure 2.11   Business intelligence framework: analytics layer

raised ad hoc or periodically by users. Business intelligence analytics covers a wide variety of subjects, and answers questions formed from historical events that are associated to predict future outcomes, so that business can profit from investment initiatives.

The analytics methods include verification analysis of which OLAP tools are a prevalent mechanism, and discovery mode analysis of which data mining tools are the prevalent mechanism. Each mode of analysis answers a specific business question or hypothesis, but the approaches are varied, and in the case of mining scenarios, the results may yield undetermined or unactionable results.

All analytical methods use association methods and business rules to act on low-granular transactional data, providing high-quality, consistent measures that are candidates for extended analysis and actionable decisions. The analytics layer provides the business analytic applications and their underlying capabilities and services, thus adding value to all areas of the organization. There are three types of analytical techniques:

Unstructured investigative: Provide a robust database of business information to analysts seeking information to support infrequent and nonrecurring business questions (modeling, mining, visualization).

Structured investigative: Deliver structured sets of information on demand to end consumers to provide answers to recurring business questions (reporting, monitoring, scorecards).

Embedded: Intelligently "push" information directly to end consumers by continuously monitoring ongoing business performance against business objectives.

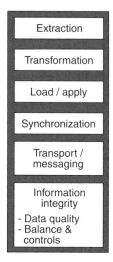

Figure 2.12   Business intelligence framework: integration layer

Data Integration Layer

The data integration layer (Figure 2.12) of the business intelligence framework defines the functions and services to source data, bring it into the warehouse operating environment, improve its quality, and format it for presentation through tools made available via the access layer.

The integration layer owns the technology and processes for moving and enhancing the data from the data sources to business intelligence repositories.

Data Source Layer

The data source layer (Figure 2.13) categorizes data as enterprise, unstructured, informational, or external. Driven by the metadata characteristics of each category, tools will be used to access and prepare the data within each category. The data source layer provides the data or raw materials (both internal and external to the organization) that will be the foundation for analysis and knowledge.

DATA ARCHITECTURE

Enterprise business intelligence can empower organizations to make their corporate data available to an unlimited number of people inside and outside the enterprise, so they can use it to make faster and more informed

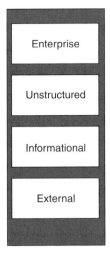

Figure 2.13   Business intelligence framework: data source layer

decisions, quickly identify new opportunities, and understand how well they're performing.

Managers and employees can track key performance indicators and obtain the insight they need to better perform their jobs, information can be shared with partners and suppliers to improve communication and collaboration, and value-added or revenue-generating information-based services can be offered to customers.

The most successful enterprise business intelligence initiatives—and those that deliver the most rapid return on investment—are those that offer maximum flexibility, accelerate deployment, and eliminate unnecessary expenditures. Enterprise business intelligence requires a powerful and flexible data architecture that will enable an organization to make all its data, regardless of its source or location, available to an increasing number of information consumers.

To make this a reality, organizations need to implement an effective data architecture strategy that will satisfy a wide array of reporting and analysis needs by supporting a variety of applications and latency requirements. It will also minimize expenditures and promote rapid return on investment by leveraging and extending the value of existing data warehouses and operational systems.

The architecture must also promote rapid return on investment and minimize the costs of developing and deploying the enterprise business intelligence environment by enabling an organization to leverage and build upon existing technology investments. Additionally, it must adapt to virtually any data warehousing or infrastructure strategy, so organizations can continue to follow the approach, or hybrid approaches, that best meet their needs.

An organization's data architecture must be optimized to support transaction-based applications, as they will often require a combination of information gathered from data warehouses, as well as operational systems. Ensuring that the proper infrastructure is in place and that all required information is readily available will help decrease deployment cycles, improve application performance, and ensure data quality, consistency, and accuracy.

As different applications, and even different usage requirements for the same application, cause different degrees of information latency and as data environments, end-user needs, and business requirements continue to change, organizations must ensure that the data architecture they implement to support their enterprise business intelligence initiatives is powerful and flexible enough to efficiently and cost-effectively satisfy their varied demands, now and in the future. A well-structured architecture will satisfy the varying information needs and latency requirements of users throughout the enterprise and beyond by making all data—whether it resides in data warehouses, data marts, or operational systems—easily accessible to anyone, anywhere, at any time.

As the amount of data an organization manages grows and data environments become more complex, the success of an enterprise business intelligence initiative becomes increasingly dependent upon the data architecture that supports it. The data architecture must be designed to meet the organization's unique information needs by:

■ Ensuring vital, relevant, and timely information is available to the right people at the right time
■ Effectively leveraging existing infrastructure and extending the value of operational systems and data warehouses
■ Minimizing management and maintenance while optimizing performance

Organizations looking for an enterprise business intelligence solution should look for an integrated suite that can scale to accommodate any number of people, any kind of reporting application, and any type of user.

Additionally, it must provide the features needed to implement and optimize the underlying data architecture such as access to any critical information asset, including databases, application systems, and transaction systems, as well as ETL capabilities, resource monitoring, and preemptive governing.

Integrated Solution

As enterprise applications continue to evolve and their data requirements change, keeping tabs on the enterprise's data assets and avoiding costly replicated efforts and errors become a challenging task. Enterprises spend significant resources keeping their data architectures manageable and useful for the people who need the assets they contain.

Over time, many of these enterprises end up consuming significant resources managing a complex analytical environment characterized by

- Maintenance of disparate software applications with differing data architectures and asynchronous release cycles
- Coordination of multiple vendor relationships
- Custom integration of each tool into the organization's overall IT infrastructure
- Continual training of user communities on the set of analytical tools

There are several methods available for integrating data assets, however, many situations are too complex or require more flexibility than some traditional options allow. Enterprises want to rationalize the complex business intelligence environment by standardizing on a single platform that maximizes benefit and minimizes cost.

The most efficient and cost-effective way to ensure that the enterprise data environment is structured to support strategic enterprise initiatives is to select a fully integrated solution that provides comprehensive reporting and business intelligence capabilities, as well as the functionality needed to implement and optimize the underlying data architecture.

Applications would benefit by leveraging existing data assets, where possible, applying integration technologies to get a coherent picture of enterprise data. Enterprises are integrating analytical tools ranging from sophisticated artificial intelligence applications (e.g., real-time fraud detection) to static HTML reports (e.g., tracking packages on a commercial shipper's Web site) into every core business functions.

In order to derive significant value from investments in analytical applications, increasingly organizations who are building or augmenting a business intelligence environment demand an integrated platform capable of delivering scalable enterprise reporting and analysis that meets a broad set of functional requirements while supporting a broad spectrum of user requirements at all levels of analytical sophistication, from statistical analysts to store clerks.

A platform approach to deploying these analytical tools must allow enterprises to derive approximately 80 percent of the needed functionality from a single business intelligence platform (e.g., data mining, OLAP, query and reporting, and report distribution) without lengthy deployment cycles and costly maintenance and training programs.

Armed with this comprehensive and integrated functionality, organizations can provide the greatest reach to enterprise data and make more information available to more people, at the lowest possible cost. Organizations that have employed the platform approach to business intelligence solutions are finding the cost and time associated with implementation and maintenance to be significantly reduced.

Enterprises should look for a solution that provides:

- The ability to access a wide range of data, including data staged in warehouses or marts and real-time data from operational systems, from any source on any platform
- Complete extract, transform, and load functionality to protect production applications and preserve system resources by reliably and cost-effectively accessing and moving data into warehouses and marts
- Full support for near real-time models, data cleansing to ensure information quality and integrity, and metadata management for complete impact analysis of the information contained in the warehouse or mart
- Tools for monitoring data access and usage patterns to provide a full understanding of what data is being accessed and how it is being used, hence enabling organizations to properly set up and optimize the architecture from the outset by helping them fully understand what information users need to secure
- The capability of continuously tracking how and when information is used over time, which will then allow for ongoing improvement of the architecture as requirements evolve and change; providing insight so that the environment can be continuously optimized for maximum performance, usability, and availability
- Preemptive query governing, so that data architects can prevent downtime and ensure consistent system performance of the reporting application and the operational systems from which it gathers data
- The ability to integrate a variety of applications to support automation of critical business processes and transactions as well as E-commerce and other service-oriented initiatives

3

BUSINESS INTELLIGENCE: STAGES

Business intelligence enables the enterprise to predict opportunities and potential problems in the business and to collaborate with suppliers in developing new and innovative products through the use of data and analytics.

Organizations typically gather information in order to assess the business environment, and cover fields such as marketing research, industry or market research, and competitor analysis. Competitive organizations accumulate business intelligence in order to gain sustainable competitive advantage, and may regard such intelligence as a valuable core competence in some instances.

Generally, Business Intelligence (BI) collectors glean their primary information from internal business sources. Such sources help decision makers understand how well they have performed. Secondary sources of information include customer needs, customer decision-making processes, the competition and competitive pressures, conditions in relevant industries, and general economic, technological, and cultural trends (see Figure 3.1). Industrial espionage may also provide business intelligence by using covert techniques. A grey area exists between "normal" business intelligence and industrial espionage.

Each business intelligence system has a specific goal, which derives from an organizational goal or from a vision statement (see Table 3.1). Both short-term and long-term goals exist. Irrespective of all its different descriptions, BI consists of three essential stages:

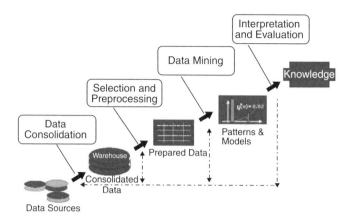

Figure 3.1   BI stages

- Move data
  - Transform, enrich, document, generate
- Store data
  - Consolidate, centralize, secure
- Access data
  - LOB ad hoc, enterprise reporting, Web portal

Table 3.1   Enterprise Business Needs and Associated Technologies

Need	Technology
Catalog data assets, track change, publish	Enterprise metadata repository
Common business language	Ontology model (including rules) for customer, items, orders, etc.
Attach objective business meaning to data	Mapping of data schemas to ontology
Manage information	Information management services: discovery, impact analysis, redundancy, classification
Integrate information	Information integration services: automatically infer and maintain data transformations and query
Virtual enterprise database	Federated query

MOVE DATA

The first step toward gaining business intelligence is to start with a "clean" database. Incomplete and inaccurate data invariably translate into incorrect management decisions. Duplicate data is also a problem as it can wrongly weigh management decisions to one side. Although a good quality database does not automatically lead to intelligent management decision making, it is a prerequisite for all types of analysis that attempt to elicit intelligent management. We could draw an analogy with cooking, where starting with the right ingredients does not guarantee you will bake a good cake, but there is very little chance you will bake a good cake if you start with the wrong set of ingredients.

One of the primary reasons enterprises do not fully realize the potential competitive advantages they can gain from their own databases is the lack of proper integration of datasets across departments. Even though all the information might reside within the organization, it may remain elusive due to a fragmentation of the data across incompatible databases. Regrouping all internal data into a single dataset or a series of interconnected datasets could be the single most useful step an organization might take toward providing a solid foundation on which quality business intelligence can be developed.

In some cases, data entry errors or missing data can also severely impair the quality of information that can be derived from corporate databases. Sorting these issues can range from very straightforward fixes (e.g., matching one list against another) to more time-consuming processes (e.g., contacting all client enterprises to update contact details of individuals working there). Ideally, all inaccuracies should be weeded out of the databases. However, limited time and monetary constraints dictate that you should bear in mind how this database will be used. The level of accuracy required will vary greatly depending on the expected use for that data.

Data cleansing and database integration can provide significant advantages for an organization over the medium to long term. However, they are both extremely time-consuming activities and can create a significant strain on internal resources, making them difficult for an organization to justify. Hiring a third party to do this job is often the best solution, allowing valuable information to be gained, without disrupting day-to-day business activities.

EXTRACT, TRANSFORM, AND LOAD (ETL)

Extract, Transform, and Load (ETL) is a process in data warehousing that involves extracting data from outside sources, transforming it to fit business needs, and ultimately loading it into the data warehouse.

ETL is important, as it is the way data actually gets loaded into the warehouse. The term ETL refers to a process that loads any database.

Extract

The first part of an ETL process is to extract the data from the source systems. Most data warehousing projects consolidate data from different source systems. Each separate system may also use a different data organization or format. Common data source formats are relational databases and flat files, but may include nonrelational database structures such as IMS or other data structures such as VSAM or ISAM. Extraction converts the data into a format for transformation processing.

Transform

The transform phase applies a series of rules or functions to the extracted data to derive the data to be loaded. Some data sources will require very little manipulation of data. However, in other cases any combination of the following transformation types may be required.

■ Selecting only certain columns to load (or if you prefer, null columns not to load)
■ Translating coded values (e.g., if the source system stores M for male and F for female but the warehouse stores 1 for male and 2 for female)
■ Encoding free-form values (e.g., mapping "Male" and "M" and "Mr" onto 1)
■ Deriving a new calculated value (e.g., sale_amount_qty * unit_price)
■ Joining data from multiple sources (e.g., lookup, merge, etc.)
■ Summarizing multiple rows of data (e.g., total sales for each region)
■ Generating surrogate key values
■ Transposing or pivoting (turning multiple columns into multiple rows or vice versa)

Load

The load phase loads the data into the data warehouse. Depending on the requirements of the organization, this process ranges widely. Some data warehouses merely overwrite old information with new data. More complex systems can maintain a history and audit trail of all changes to the data.

Challenges

ETL processes can be quite complex, and significant operational problems can occur with improperly designed ETL systems.

The range of data values or data quality in an operational system may be outside the expectations of designers at the time validation and transfomation rules are specified. Data profiling of a source during data analysis is recommended to identify the data6767 conditions that will need to be managed by transform rules specifications.

The scalability of an ETL system across the lifetime of its usage needs to be established during analysis. This includes understanding the volumes of data that will have to be processed within service level agreements. The time available to extract from source systems may change, which may mean the same amount of data may have to be processed in less time. Some ETL systems have to scale to process terabytes of data to update data warehouses with tens of terabytes of data. Increasing volumes of data may require designs that can scale from daily batch to intraday microbatch to integration with message queues for continuous transformation and update.

An additional difficulty is making sure the data being uploaded is relatively consistent. Because multiple source databases all have different update cycles (some may be updated every few minutes, whereas others may take days or weeks), an ETL system may be required to hold back certain data until all sources are synchronized. Likewise, where a warehouse may have to be reconciled to contents in a source system or with the general ledger, establishing synchronization and reconciliation points is necessary.

Tools

Although ETL processes can be created using almost any programming language, creating them from scratch is quite complex. Increasingly, enterprises are buying ETL tools to help in the creation of ETL processes.

A good ETL tool must be able to communicate with the many different relational databases and read the various file formats used throughout an organization. ETL tools have started to migrate into enterprise application integration, or even enterprise service bus, systems that now cover much more than just the extraction, transformation, and loading of data. Many ETL vendors now have data profiling, data quality, and metadata capabilities.

Store Data

Throughout the history of systems development, the primary emphasis had been given to the operational systems and the data they process. However, it is not practical to keep data in the operational systems indefinitely, and only as an afterthought was a structure designed for archiving the data that the operational system has processed.

The primary concept of data storage is that the data stored for business analysis can most effectively be accessed by separating it from the data in the operational systems (see Figure 3.2). Many of the reasons for this separation have evolved over the years. The most important reason for separating data for business analysis from the operational data has always

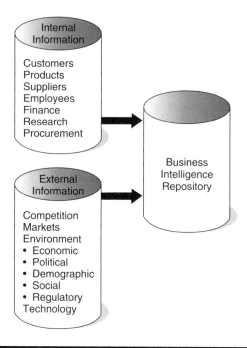

Figure 3.2 BI repository

been the potential performance degradation on the operational system that can result from the analysis processes.

High performance and quick response time is almost universally critical for operational systems. The loss of efficiency and the costs incurred with slower responses on the predefined transactions are usually easy to calculate and measure. For example, a loss of five seconds of processing time is perhaps negligible in and of itself, but it compounds out to considerably more time and high costs once all the other operations it affects are brought into the picture. On the other hand, business analysis processes in a data warehouse are difficult to predefine and they rarely need to have rigid response time requirements.

This difficulty in accessing operational data is amplified by the fact that many operational systems are often 10 to 15 years old. The age of some of these systems means that the data access technology available to obtain operational data is itself dated. In the past, legacy systems archived data onto tapes as it became inactive and many analysis reports ran from these tapes or mirror data sources to minimize the performance impact on the operational systems.

These reasons to separate the operational data from analysis data have not significantly changed with the evolution of data warehousing systems,

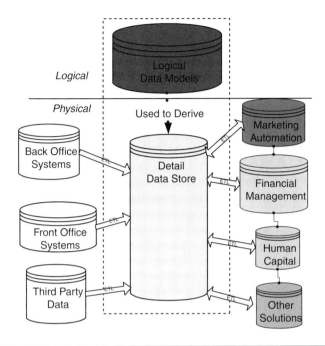

Figure 3.3   Database model

except that now they are considered more formally during the data warehouse building process. Advances in technology and changes in the nature of business have made many of the business analysis processes much more complex and sophisticated. In addition to producing standard reports, today's data warehousing systems support very sophisticated online analysis including multidimensional analysis (see Figure 3.3).

Operational systems process data to support critical operational needs. In order to do that, operational databases historically have been created to provide an efficient processing structure for a relatively small number of well-defined business transactions. However, because of the limited focus of operational systems, the databases designed to support operational systems have difficulty accessing the data for other management or informational purposes.

Clearly, the goal of data storage systems such as the data warehouse is to free the information that is locked up in the operational databases and to mix it with information from other, often external, sources of data. Increasingly, large organizations are acquiring additional data from outside databases. This information includes demographic, econometric, competitive, and purchasing trends. The so-called "information superhighway" is providing access to more data resources every day.

DATA WAREHOUSE

Data warehousing is not a new phenomenon. All large organizations already have data warehouses, but they are just not managing them. Over the next few years, the growth of data warehousing is going to be enormous with new products and technologies coming out frequently.

Data warehousing has quickly evolved into a unique and popular business application class. Numerous examples can be cited of highly successful data warehouses developed and deployed for businesses of all sizes and all types. Hardware and software vendors have quickly developed products and services that specifically target the data warehousing market.

More and more enterprises are using data warehousing as a strategy tool to help them win new customers, develop new products, and lower costs. Searching through mountains of data generated by corporate transaction systems can provide insights and highlight critical facts that can significantly improve business performance.

Until recently, data warehousing has been an option mostly for big enterprises, but the reduced costs of warehousing technology make it a practical, often even a competitive, requirement for smaller enterprises as well. Turnkey integrated analytical solutions are reducing the cost, time, and risk involved in data warehouse implementations.

Although access to the warehouse was previously limited to highly trained analytical specialists, corporate portals now make it possible to grant data access to hundreds or thousands of employees.

What Is a Data Warehouse

The simplest description would be that a data warehouse is managed data situated after and outside the operational systems.

However, a data warehouse is far more complex than it looks. It is a logical collection of information gathered from many different operational databases used to create business intelligence that supports business analysis activities and decision-making tasks, primarily, a record of an enterprise's past transactional and operational information, stored in a database designed to favor efficient data analysis and reporting (especially OLAP).

A data warehouse can be considered a copy of transaction data specifically structured for querying and reporting. However, sometimes non-transaction data are stored in a data warehouse, although probably 95 to 99 percent of the data usually are transaction data.

The data in a data warehouse system is brought to it after it has become mostly nonvolatile. This means that after the data is in the data warehouse, there are no modifications to be made to this information. For example, the order status does not change, the inventory snapshot does not change, and the marketing promotion details do not change. This attribute of the data warehouse has many very important implications for the kind of data that is brought to the data warehouse and the timing of the data transfer.

A data warehouse can be normalized or denormalized. It can be a relational database, multidimensional database, flat file, hierarchical database, object database, and so on. Data warehouse data often gets changed, and data warehouses often focus on a specific activity or entity.

Some of the activity against today's data warehouses is predefined and not much different from traditional analysis activity. However, other processes such as multidimensional analysis and information visualization were not available with traditional analysis tools and methods.

Data warehouses touch the organization at all levels, and the people that design and build the data warehouse must work across the organization. The industry and product experience of a diverse team, coupled with a business focus and proven methodology, can make a huge difference.

With the average cost of a data warehousing system valued at more than 1 million U.S. dollars, the right people, methodology, and experience are critical. The reliance on technology is only a small part of realizing the true business value buried within the multitude of data collected within organizations today.

Data Warehouse Architecture

A Data Warehouse Architecture (DWA; see Figure 3.4) is a way of representing the overall structure of data, communication, processing, and presentation that exists for end-user computing within the enterprise. The architecture is made up of a number of interconnected parts:

- *Operational Database/External Database Layer:* The external database layer of the data warehouse architecture is the layer that deals with the external operational database.
- *Information Access Layer:* The information access layer of the data warehouse architecture is the layer with which the end user deals directly.
- *Data Access Layer:* The data access layer of the data warehouse architecture is involved with allowing the information access layer to talk to the operational layer.
- *Data Directory (Metadata) Layer:* The data directory (metadata) layer is involved in provision of universal data access as it interacts with the data directory or repository of metadata information.
- *Process Management Layer:* The process management layer is involved in scheduling the various tasks that must be accomplished to build and maintain the data warehouse and data directory information.
- *Application Messaging Layer:* The application message layer has to do with transporting information around the enterprise computing

network. Application messaging is also referred to as "middleware," but it can involve more than just networking protocols.

■ *Data Warehouse (Physical) Layer:* The (core) data warehouse is where the actual data utilized primarily for informational uses occurs. In some cases, one can think of the data warehouse simply as a logical or virtual view of data.

■ *Data Staging Layer:* The final component of the data warehouse architecture is data staging; it includes all the processes necessary to select, edit, summarize, combine, and load the data warehouse and access data from operational and external databases.

Design of Data Warehouses

In order to get the most from a data warehouse, it is important that data warehouse planners and developers have a clear idea of what they are looking for and then choose strategies and methods that will provide them with performance today and flexibility for tomorrow.

Data warehouses often hold large amounts of information that are sometimes subdivided into smaller logical units called dependent data marts. Dependent data marts allow for easier reporting by keeping relevant data together in one location.

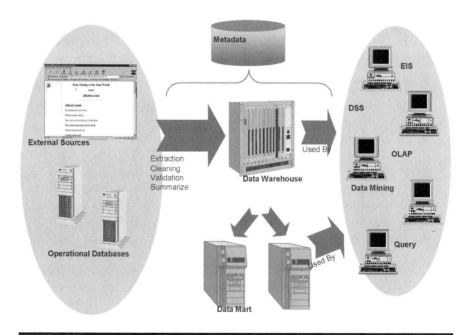

Figure 3.4   Data warehouse architecture

Metadata	Definition	Transformation/ Derivation	Management/ Administration
Business (mostly unstructured)	• What does it mean? • Where can I find it?	• How was it calculated? • What were the sources? • What business rules were applied?	• What training is available? • Who's on the steering team? • What's the easiest way to get to it? • How fresh is the information?
Technical (mostly structured)	• format • length • domain • database • catalog	• filters • aggregates • calculations • expressions	• capacity planning • space allocation • indexing & reindexing • disk utilization • production job scheduling

Figure 3.5   Structured and unstructured data

Usually, two basic ideas guide the creation of a data warehouse:

■ Integration of data from distributed and differently structured data-bases, which facilitates a global overview and comprehensive analysis in the data warehouse
■ Separation of data used in daily operations from data used in the data warehouse for purposes of reporting, decision support, analysis, and controlling

Periodically, one imports data from Enterprise Resource Planning (ERP) systems and other related business software systems into the data warehouse for further processing. It is common practice to stage data prior to merging it into a data warehouse. In this sense, to "stage data" means to queue it for preprocessing, usually with an ETL tool. The preprocessing program reads the staged data (often a business's primary OLTP databases), performs qualitative preprocessing or filtering (including denormalization, if deemed necessary), and writes it into the warehouse.

Dimensions and Measures

A data warehouse is created by analyzing ways to categorize data using dimensions and ways to summarize data using measures. Dimensions can be used to filter data by excluding results or by displaying data in different cells of a presentation. Measures are used to create averages and totals using recomputed aggregates.

Data Warehouse Implementation Methods

There are numerous ways to build and deploy a data warehouse environment, but a few primary methods have emerged (see Figure 3.6).

Top-Down Approach

The top-down approach calls for a single, centralized data warehouse containing both summary and detailed data and smaller dependent data marts that derive all their data from the data warehouse.

The Bottom-Up Approach

The goal of the bottom-up approach is to "deliver business value by deploying dimensional data marts as quickly as possible." This approach is both flexible and user-friendly. And, because heavy infrastructure is not required at the outset, value is delivered quickly.

The Federated Approach

The federated approach is not a defined architecture, but a theory that allows for any means necessary to integrate data assets in order to meet

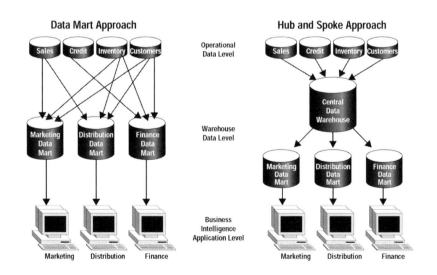

Figure 3.6   Data warehouse approaches

emerging needs and respond to dynamic conditions. The federated approach encourages organizations to share the "highest value" metrics, dimensions, and measures, wherever possible and however possible. This approach, according to some of its supporters, provides the maximum amount of architecture possible in a given political and implementation reality.

The Need for Staged Data

The fundamental requirements of the operational and analysis systems are different: the operational systems need performance, whereas the analysis systems need flexibility and broad scope. It has rarely been acceptable to have business analysis interfere with and degrade performance of the operational systems. Today's most popular and widely implemented business intelligence strategies involve building a data warehouse or data marts as the centerpiece of the business intelligence infrastructure.

Generally, data warehousing is not meant for current "live" data, although "virtual" or "point-to-point" data warehouses can access operational data. A "real" data warehouse is generally preferred to a virtual DW because stored data has been validated and is set up to provide reliable results to common types of queries used in a business. Although the data warehouse keeps validated data, data marts serve operational requirements. The creation and deployment of data warehouses and data marts serve as a means of:

- Facilitating an aggregated or holistic view of information from multiple systems
- Improving the availability of data to business users and simplifying access by masking the underlying complexity of operational data structures
- Eliminating potential performance drains on operational systems

Integrating Data from Multiple Operating Systems

Data warehousing systems are most successful when data can be combined from more than one operational system. When the data needs to be brought together from more than one source application, it is natural that this integration is done at a place independent of the source applications. The data warehouse may very effectively combine data from multiple source applications such as sales, marketing, finance, and production. Many large data warehouse architectures allow for the source applications to be integrated into the data warehouse incrementally.

The primary reason for combining data from multiple source applications is the ability to cross-reference data from these applications. Nearly all data in a typical data warehouse is built around the time dimension. Time is the primary filtering criterion for a very large percentage of all activity against the data warehouse.

The time dimension in the data warehouse also serves as a fundamental cross-referencing attribute. The ability to establish and understand the correlation between activities of different organizational groups within an organization is often cited as the single biggest advanced feature of the data warehousing systems.

The data warehouse system can serve not only as an effective platform to merge data from multiple current applications but it can also integrate multiple versions of the same application. The data warehouse system can serve as a very powerful and much needed platform to combine the data from the old and the new applications. Designed properly, the data warehouse can allow for year-on-year analysis even though the base operational application has changed.

DATA MARTS

A data mart is a specialized version of a data warehouse. Like data warehouses, data marts contain a snapshot of operational data that helps business people to strategize based on analyses of past trends and experiences.

The key difference is that the creation of a data mart is predicated on a specific, predefined need for a certain grouping and configuration of select data. Because data mart configuration emphasizes easy access to relevant information, the star schema is a fairly popular design choice, as it enables a relational database to emulate the analytical functionality of a multidimensional database. At the same time trade-offs inherent with data marts include limited scalability, duplication of data, data inconsistency with other silos of information, and inability to leverage enterprise sources of data.

A dependent data mart is a logical subset (view) or a physical subset (extract) of a larger data warehouse, isolated for one of the following reasons:

■ A need for a special data model or schema: for example, to restructure for OLAP
■ To offload the data mart to a separate computer for greater efficiency or to obviate the need to manage that workload on the centralized data warehouse
■ To separate an authorized data subset selectively
■ To bypass the data governance and authorizations required to incorporate a new application on the enterprise data warehouse
■ To demonstrate the viability and ROI (Return On Investment) potential of an application prior to migrating it to the enterprise data warehouse

- A coping strategy for IT (Information Technology) in situations where a user group has more influence than funding or is not a good citizen in the centralized data warehouse
- A coping strategy for consumers of data in situations where a data warehouse team is inept and unable to create a usable data warehouse

OLAP

Enterprises have been storing multidimensional data, using a star or snowflake schema, in relational databases for many years. Over time, relational database vendors have added optimizations that enhance query performance on these schemas. During the 1990s many special-purpose databases were developed that could handle added calculation complexity and that generally performed better than relational engines, an example being DB2, which has added a number of features that make it more competitive with these special-purpose databases.

Online analytical processing is a term that was introduced as a play on the previously familiar term OnLine Transaction Processing (OLTP) and signaled a shift in the paradigm for business analysis, in parallel with the shift that had already occurred for transaction processing. Instead of reviewing piles of static reports printed on green-bar paper, the OLAP analyst could explore business results interactively, dynamically adjusting the view of the data, asking questions and getting answers almost immediately.

This freedom from static answers to fixed questions on a fixed schedule allows business analysts to operate more effectively and to effect improvements in business operations. The ultimate meaning of OLAP is left to time and the growing collection of products that claim to be OLAP products, or have OLAP components.

What Is OLAP?

OLAP is an approach to quickly provide the answer to analytical queries that are dimensional in nature. It is part of the broader category of business intelligence, which also includes ETL, relational reporting, and data mining. The concept of OLAP can also be described as the Fast Analysis of Shared Multidimensional Information (FASMI). It borrows aspects of navigational databases and hierarchical databases that are speedier than their relational kin. Databases configured for OLAP employ a multidimensional data model, allowing for complex analytical and ad hoc queries with a rapid execution time.

- *Fast:* Means that the system is targeted to deliver the most responses to users within about 5 seconds, with the simplest analyses taking no more than 1 second and very few taking more than 20 seconds.

- *Analysis:* Means that the system can cope with any business logic and statistical analysis that is relevant for the application and the user, and keep it easy enough for the target user.
- *Shared:* Means that the system implements all the security requirements for confidentiality and, if multiple write access is needed, concurrent update locking at an appropriate level.
- *Multidimensional:* Means that the system must provide a multidimensional conceptual view of the data, including full support for hierarchies and multiple hierarchies, as this is certainly the most logical way to analyze businesses and organizations.
- *Information:* Includes all of the data and derived information needed, wherever it is and however much is relevant for the application.

OLAP takes a snapshot of a set of source data and restructures it into an OLAP cube. The queries can then be run against this. It has been claimed that for complex queries OLAP can produce an answer in around 0.1 percent of the time for the same query on OLTP relational data.

The cube is created from a star schema of tables. At the center is the fact table which lists the core facts that make up the query. Numerous dimension tables are linked to the fact tables. These tables indicate how the aggregations of relational data can be analyzed. The number of possible aggregations is determined by every possible manner in which the original data can be hierarchically linked.

The calculation of the aggregations and the base data combined make up an OLAP cube, which can potentially contain all the answers to every query that can be answered from the data. Due to the potentially large number of aggregations to be calculated, often only a predetermined number are fully calculated and the remainder are solved on demand.

Unlike relational databases—which had SQL as the standard query language and widespread APIs such as ODBC, JDBC, and OLEDB—there long was no such unification in the OLAP world. The first real standard API was OLEDB for OLAP specification from Microsoft which appeared in 1997 and introduced the MDX query language. Several OLAP vendors, both server and client, adopted it. In 2001 Microsoft and Hyperion announced the XML for Analysis specification, which was endorsed by most of the OLAP vendors. Because this also used MDX as a query language, MDX became the de facto standard in the OLAP world.

Database explosion is a phenomenon causing a vast amount of storage space being used by OLAP databases when certain but frequent conditions are met: high number of dimensions, precalculated results, and sparse multidimensional data. The difficulty in implementing OLAP comes in forming the queries, choosing the base data and developing the schema, as a result of which most modern OLAP products come with huge libraries of preconfigured queries. Another problem is in the base data quality: it must be complete and consistent.

The typical applications of OLAP are in business reporting for sales, marketing, management reporting, Business Performance Management (BPM), budgeting and forecasting, financial reporting, and similar areas.

OLAP systems, first described by E.F. Codd in a white paper, "Providing OLAP to User Analysts: An IT Mandate," as a slight modification of the traditional database term OLTP, have multidimensionality as the primary requirement. OLAP products present their data in a multidimensional framework. Dimensions are collections of related attributes (product, market, time, or customer) called identifiers, of the data values of the system.

The identifiers belonging to the collection for a particular dimension generally have some sort of structure, usually hierarchical. Sometimes there is more than one natural structure for these identifiers. The multidimensional characteristic means that an OLAP system can quickly switch among various orientations of dimensions, as well as among various subsets and structural arrangements of a dimension. Because of the multidimensional nature of OLAP systems, the collections of data that they implement are often referred to as cubes.

An OLAP Example

To give a feeling of how one should see OLAP, let us look at the following simple example. Consider a shoe retailer with many shops in different cities and many different styles of shoes, for example, ski boot, gumboot, and sneaker, as shown in Figure 3.7. Each shop delivers data daily on quantities sold in numbers per style. This data is stored centrally. Now the business analyst wants to follow sales by month, outlet, and style. These are called dimensions, for example, month dimension. If we want to look at the data of these three dimensions and say something significant about them, what we are actually doing is looking at the data stored in a three-dimensional cube.

The three cubes in the second line show us how we can look at data on all shoe styles sold in all months in the outlet Amsterdam, data on shoe style sneaker sold in all months in all outlets, and data on all shoe styles sold in all outlets in the month of April.

When we combine these three dimensions, we get data on the number of sneakers sold in the outlet Amsterdam in the month of April.

Suppose we want information about the colors of the sneakers or the sizes sold: we would have to define new dimensions. This would mean a four-, five-, or even more-dimensional cube. Of course cubes such as these are no longer "visible" to the eye, but in an OLAP-application they are possible.

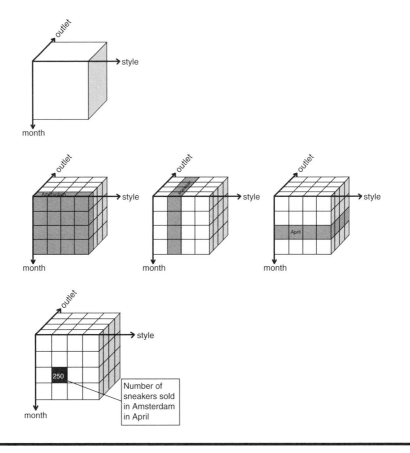

Figure 3.7 OLAP example

Types of OLAP

There are three types of OLAP. Each type has certain benefits, although there is disagreement about the specifics of the benefits between providers.

Multidimensional OLAP (MOLAP)

MOLAP is the classic form of OLAP and is sometimes referred to as just OLAP. MOLAP uses database structures that are generally optimized for retrieval, for example, arrays or compressed records. Unlike a relational database, these storage forms are optimized for speed of calculation. Often they are also optimized for retrieval along hierarchical access patterns. The dimensions of each cube are typically attributes such as

Table 3.2   OLAP Characteristics

Fast

To keep the spirit of the "O–On-Line" in OLAP, systems need to provide results very quickly, in just a few seconds. This level of performance is central to allowing analysts to work effectively without distraction.

Analytic

To keep the spirit of the "A–Analytical" in OLAP, systems need to provide rich analytic functions appropriate to a given application, with minimal programming.

Shared

OLAP systems are usually shared resources that have a security requirement and integrity features. Ultimately, this can mean providing different access controls on each cell of a database.

Multidimensional

OLAP products present their data in a multidimensional framework where dimensions are collections of related attributes (product, market, time) of the data values of the system.

---

time period, location, and product or account code. The way that each dimension will be aggregated is defined in advance by one or more hierarchies.

MOLAP is better on smaller sets of data as it is faster to calculate the aggregations and return answers and does need less storage space.

Relational OLAP (ROLAP)

ROLAP works directly with relational databases; the base data and the dimension tables are stored as relational tables and new tables are created to hold the aggregated information.

ROLAP is considered more scalable. However, large volume preprocessing is difficult to implement efficiently so it is frequently skipped. ROLAP query performance can therefore suffer.

Hybrid OLAP (HOLAP)

There is no clear agreement across the industry as to what constitutes hybrid OLAP, except that a database will divide data between relational and specialized storage. For example, for some vendors, a HOLAP database will use relational tables to hold the larger quantities of detailed data, and use specialized storage for at least some aspects of the smaller quantities of more-aggregate or less-detailed data.

HOLAP is between the other two in all areas, but it can preprocess quickly and scale well. All types though are prone to database explosion.

BALANCED SCORECARD

A product that many BI companies have to offer is the balanced scorecard. The Balanced ScoreCard (BSC) was born in 1992, when Robert Kaplan and David Norton published an article about it in the *Harvard Business Review*.

The balanced scorecard application is the marriage of the balanced scorecard methodology and advanced OLAP technology. Apparently the applications based on the BSC methodology are often integrated with OLAP environments. The BSC is a simple control mechanism that helps managers to monitor their business performance in the following four perspectives:

1. Customer knowledge
2. Financial performance
3. Internal business processes
4. Learning and growth

All BSC tools are able to lay down Critical Success Factors (CSFs) in the four perspectives named above. These factors can be connected to the Key Performance Indicators (KPIs). A good BSC contains all the KPIs that are critical for achieving the company's strategic goals. Based on the KPIs, decisions for the appropriate actions can be made.

The use of BSC revolves around:

■ What do we want to achieve with the organization? (strategic goals)
■ In order to achieve our goals, what should we be good at? (critical success factors)
■ How can we measure that we are achieving what we want? (key performance indicators)

The BSC is focused on the internal management of an organization, whereas most other BI applications are focused on external business. The BSC is a tool for managers, and other BI tools are suitable for all organizational levels (see Table 3.3).

Table 3.3 OLAP Applications

Application Area	Description
Marketing and sales analysis	Mostly found in consumer goods industries, retailers, and the financial services industry.
Database marketing	Determine who the best customers are for targeted promotions for particular products or services.
Financial reporting	To address this specific market, certain vendors have developed specialist products.
Management reporting	Using OLAP-based systems one is able to report faster and more flexibly, with better analysis than the alternative solutions.
Balanced scorecard	It is the marriage of the balanced scorecard methodology and advanced OLAP technology.
Profitability analysis	Important in setting prices and discounts, deciding on promotional activities, selecting areas for investment or divestment, and anticipating competitive pressures.
Quality analysis	OLAP tools provide an excellent way of measuring quality over long periods of time and of spotting disturbing trends before they become too serious.

DATA MINING

Data mining, also known as Knowledge Discovery in Databases (KDD), is the process of automatically searching large volumes of data for patterns. In order to achieve this, data mining uses computational techniques from statistics, machine learning, and pattern recognition.

Data mining is often confused with "writing reports and queries," when in fact data mining activities do not involve any traditional report writing or querying at all. Data mining is performed through a specialized tool, which executes predefined data mining operations based on analytical models. Data mining is the analysis of data with the intent to discover gems of hidden information in the vast quantity of data that has been captured in the normal course of running the business.

Defining Data Mining

There are many good definitions of the term "data mining" but the core concept behind all of them is the same.

IBM defines data mining as

> *The process of extracting previously unknown, valid and actionable information from large databases and then using the information to make crucial business decisions.*

and

> *The science of extracting useful information from large data sets or databases.*

Other definitions are not concerned with the size of the database or whether the information is used in business but, in practice, these conditions are common. So, what's meant by "previously unknown, valid and actionable information"?

The information is previously unknown in that it is not directly derived from the data. Instead, the information takes the form of relationships among the database columns where the value in one or more columns predicts the outcome in another, hence the name, predictive model. But predictive models must be valid. Rating a model's predictive power usually involves testing it against another data set. The information is actionable, which suggests that there should be some goal in mind before putting any effort into data mining (see Figure 3.8).

A simple (though hypothetical) example of data mining is its use in a retail sales department or a very large chain of supermarkets. If a store tracks the purchases of a customer and notices that a customer buys a lot of silk shirts, the data mining system will make a correlation between that customer and silk shirts, and thereby can plan promotional schemes. Similarly in the case of a supermarket, through intensive analysis of the transactions and the goods bought over a period of time, analysts found that milk and chocolate powder were often bought together. Although explaining this interrelation might be difficult, taking advantage of it, on the other hand, should not be hard (e.g., placing the high-profit chocolate powder next to the high-profit milk). This technique is often referred to as market basket analysis.

Data Mining and Statistical Analysis

Although it is usually used in relation to analysis of data, data mining, like artificial intelligence, is an umbrella term and is used with varied meanings in a wide range of contexts. It is usually associated with a business or other organizational need to identify trends. Data mining is different from conventional statistical analysis.

In statistical analyses, in which there is no underlying theoretical model, data mining is often approximated via stepwise regression methods wherein the $2k$ of possible relationships between a single outcome variable and k potential explanatory variables is smartly searched. With the advent of parallel computing, it became possible (when k is less than approximately 40) to examine all $2k$ models. This procedure is called all subsets or exhaustive regression.

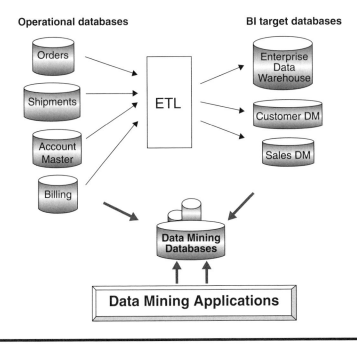

Figure 3.8 Data-mining framework

Data-Mining Operations

Data-mining tools enable statisticians to build analytical models, which the tools use during data-mining operations. The results of data-mining operations are tables and files loaded with analysis data, which can then be accessed with query and reporting tools. A predictive engine will ask for a list of input criteria, and will follow the steps and relationships from the analytical model to determine the most likely predictions. There are four common data mining operations.

1. *Predictive and classification modeling:* Predictive and classification modeling is used to forecast a particular event. It assumes that an analyst has a specific question he or she would like to ask. The model provides the answer by assigning ranks, which determines the likelihood of certain classes.
2. *Link analysis:* Link analysis finds relationships between database records.
3. *Database segmentation:* Database segmentation groups related records into segments. This grouping is often the first step of data selection, before other data-mining operations take place.

4. *Deviation detection:* Deviation detection looks for records that fall outside the norm and suggests reasons for the anomalies.

Data Mining—Data Sources

Data warehouse databases are popular sources for data-mining applications because they contain a wealth of internal data from across business boundaries, which was gathered, consolidated, validated, and cleansed in the extract/transform/load process. DW databases may also contain valuable external data, such as regulations, demographic, or geographic data, which when combined with internal organizational data offer a firm foundation for data mining (see Table 3.4).

However, once the data has been summarized for the DW, hidden data patterns, data relationships, and data associations are often no longer discernable from that data pool. For example, the tool may not be able to perform the common data-mining task of market basket analysis with sales data, which was summarized by week, by product, and by store because some detailed data pattern about each sale may have gotten lost in the summary. Therefore, operational files and databases are also popular sources for data-mining applications because they contain transaction-level detailed

Table 3.4   Statistical Analysis Versus Data Mining

Statistical Analysis	*Data Mining*
Statisticians usually start with a hypothesis (a question or assumption).	Data mining does not require a hypothesis.
Statisticians have to develop their own equations to match their hypotheses.	Data-mining algorithms in the tool can automatically develop the equations.
Statistical analysis uses only numerical data.	Data-mining tools can use different types of data, not just numerical data.
Statisticians can find and filter dirty data during their analysis.	Data mining depends on clean, well-documented data.
Statisticians interpret their own results and convey these results to business managers and executives.	Data-mining results are not easy to interpret, and a statistician must still be involved in analyzing the data-mining results and conveying the findings to the business managers and executives.

data with myriad hidden data patterns, data relationships, and data associations. Data-mining tools can theoretically access the operational databases and DW databases directly without building data-mining databases first, as long as the database structures are supported by the tool (e.g., relational such as Oracle).

However, accessing operational and DW databases directly is not an advisable practice for several reasons:

- The data pool needs to be able to change for the data-mining tool, such as dropping a sales region or restricting a product type for specific mining purposes. Changing the data content of operational or DW databases is not possible.
- The performance of operational as well as DW databases would be affected by the data-mining operations. That is unacceptable for operational databases, and not desirable for DW databases.
- A data-mining operation may need detailed historical data. Operational databases do not store historical data and DW databases often do not have the desired level of detail.

It is therefore common that organizations extract data for data mining from their DW databases and from their operational files and databases as needed into special-purpose data-mining databases (see Table 3.5).

Table 3.5   Steps of Data-Mining Process

Step in the Process	Description
Business understanding	Determining the business objectives, situation assessment, determining the goal of the data mining, producing a project plan
Data understanding	Collecting the initial data, describing and exploring this data, and verifying its quality
Data preparation	Selecting, cleaning, constructing, integrating, and formatting the data
Modeling	Selecting a modeling technique, generating test design, building and implementing the model
Evaluation	Evaluating the results, reviewing the process, and determining the next steps
Deployment	Plan deployment, plan monitoring and maintenance, producing the final report, and reviewing the project

Why Use Data Mining?

First, most organizations are sitting on top of a gold mine, the "gold" being all the data collected about their customers and the products their customers buy. Embedded in this data are their customers' styles of expenditure, their likes and dislikes, and other information about their buying habits. It is a wasted resource not to use this business intelligence hidden in the data.

Second, an increasingly competitive business environment drives the need for data mining. For instance, the competition between telephone enterprises has created the need for "churn" analysis. A telephone organization that doesn't act to keep customers and attract new ones will not survive long. Data mining uses algorithms to sift through huge volumes of information for the purpose of detecting patterns hidden in the data. Understanding these patterns quickly leads to improved business intelligence.

Data Dredging

Used in the technical context of data warehousing and analysis, the term "data mining" is neutral. However, it sometimes has a more depreciatory usage that implies imposing patterns (and particularly causal relationships) on data where none exist. This imposition of irrelevant, misleading, or trivial attribute correlation is more properly criticized as "data dredging" in the statistical literature. Another term for this misuse of statistics is data fishing. Used in this latter sense, data dredging implies scanning the data for any relationships, and then when one is found coming up with an interesting explanation. (This is also referred to as "overfitting the model.")

The problem is that large data sets invariably happen to have some exciting relationships peculiar to that data. Therefore any conclusions reached are likely to be highly suspect. In spite of this, some exploratory data work is always required in any applied statistical analysis to get a feel for the data; therefore, sometimes the line between good statistical practice and data dredging is less than clear.

The common approach, in data mining, to overcoming the problem of overfitting is to separate the data into two or three separate data sets (called the training set, validation set, and testing set). The model is built using the training and validation set, and is then tested using the testing set; the procedure can be repeated many times by resampling the data sets, in order to be more certain that a real pattern has been found and that the model is not merely capitalizing on random chance (i.e., overfitting).

A more significant danger is finding correlations that do not really exist. "There have always been a considerable number of people who busy themselves examining the last thousand numbers which have appeared on a roulette wheel, in search of some repeating pattern. Sadly enough, they have usually found it."

However, when properly done, determining correlations in investment analysis has proven to be very profitable for statistical arbitrage operations (such as pairs trading strategies), and furthermore correlation analysis has shown to be very useful in risk management.

Most data-mining efforts are focused on developing a finely grained, highly detailed model of some large dataset. Other researchers have described an alternate method that involves finding the minimal differences between elements in a data set, with the goal of developing simpler models that represent relevant data.

Data-Mining Techniques

Analyzing the information that an enterprise stores in connection with all customer interactions can reveal a lot of remarkable facts about the buying behavior of customers, what motivates them, and what might make them buy from competitors. It also provides a scientific method to monitor enterprise business performance.

Data-mining techniques are specific implementations of algorithms that are used in data-mining operations. When deciding to mine information from a database, one is faced with a wide number of available techniques. Basic statistical measurements such as means, variances, and correlation coefficients are useful in the early stages of data analysis to gain an overall view of the structure of the data. By revealing simple interrelations within the data, statistical modeling can show which in-depth technique is likely to bring further information relevant to the current task. Some of the more popular data-mining techniques are:

- *Association:* Association analysis is used to identify the behavior of specific events or processes. Associations link occurrences within a single event. Association analysis is sometimes called market basket analysis.
- *Sequence:* Sequences are similar to associations, but they link events over time and determine how items relate to each other over time.
- *Classification:* Classification is the most common use of data mining. Classifications look at the behavior and attributes of already determined groups. The groups might include frequent flyers, high spenders, loyal customers, people responding to direct mail campaigns, or people with frequent back problems (people driving long distances every day). The data-mining tool can assign classifications to new data by examining existing data that has already been classified and by using the results to infer a set of rules. The set of rules is then applied to any new data to be classified. This technique often uses supervised induction, which employs

a small training set of already classified records to determine additional classes.

■ *Cluster:* Clustering is a technique that aggregates data according to a predetermined set of characteristics. It can be used to differentiate groups of customers that behave similarly on certain factors, within the data. This is similar to classification, except that no groups have yet been defined at the outset of running the data-mining tool. The clustering technique often uses neural networks or statistical methods. Clustering divides items into groups based on the similarities the data-mining tool finds. Within a cluster the members should be very similar, but the clusters themselves should be very dissimilar. Clustering is used for problems such as detecting manufacturing defects or finding affinity groups for credit cards.

■ *CHAID Analysis:* CHAID analysis, which stands for CHi-square Automatic Interaction Detection, can be seen as the opposite of clustering, in the sense that the CHAID analysis starts with the overall database, and then splits it according to the most important variable until it achieves homogeneous subgroups that cannot be split any further. A major advantage of this technique is that the results can be presented as an easy-to-read classification tree, with each split in the tree being accredited to a single variable (e.g., credit worthiness, income, age, etc.).

■ *Regression (Forecasting):* Regression is one of two forecasting techniques. It uses known values of data to predict future values or future events based on historical trends and statistics.

■ *Time Series (Forecasting):* The difference between regression and time series is that time series forecast only time-dependent data values. The property of time can also include a hierarchy of time periods, such as work week versus calendar week, holidays, seasons, or date ranges and date intervals.

■ *Propensity Models (Forecasting):* Propensity models also known as predictive models have proven to be very valuable in predicting which customers are most likely to purchase a certain product based on a set of current customers. The results of such a model can be directly used to develop more appropriately targeted marketing campaigns.

Other recognized techniques to extract information from datasets are database segmentation, neural networking, and wavelet analysis among others.

It can be intimidating to choose which method will provide the best results; analysis tools can differ greatly in their approach to the problem. It is therefore very important for an organization to consult someone with extensive experience in data-mining processes before going ahead with a business intelligence project. The best method to use will vary greatly depending on the time available to do the analysis, what the results will be used for, and the type of data that is available for the analysis.

An important point to consider is whether your analysis is guided by predefined questions. Predefined points of analysis are aimed at understanding certain types of behaviors by analyzing relationships among various predecided influencing factors. The techniques used to analyze data are complex. In order to be able to use the results of the data analysis, it is crucial that the results should not be clouded by the complexity of the calculations but are delivered in a straightforward manner.

Also important is to keep in mind that in spite of all the dazzling technologies, data mining has to be driven by strong business needs in order to justify the expenditure in time and money.

DATA MANAGEMENT

Today organizations must build new systems, implement new strategies, or identify new markets in order to compete or survive. What has historically been ignored is proper management of the data that supports organizations' ability to make reasonable, results-oriented decisions. Organizations misunderstand how effective data management yields a competitive advantage. Put simply, present-day organizations depend on data like never before.

Almost all the organizations know that data is an important corporate asset. But data is quite different from other corporate assets as it is the only business resource that is completely reusable. Regardless of industry, revenue size, or competitive environment, every organization relies on its data to produce information that can guide effective decisions. This data can be used by organizations any number of times, for any number of different purposes, by any number of departments simultaneously, and still it remains available for use again.

However, unlike tangible corporate assets which have a structured value, depreciation schedule, and so forth, it is difficult for many enterprises to place a definitive value on data. As a result, the perceived lack of a tangible value makes justifying data management efforts a tricky endeavor.

In recent years, forward-thinking enterprises have begun to understand one key idea: the quality of the results from any analysis is only as good as the quality of the input (the data) that feeds that analysis, hence the cost of ineffective data management is much higher than the cost of successfully managing data.

Data management establishes and deploys the roles, responsibilities, policies, and procedures pertaining to the acquisition, maintenance, dissemination, and disposition of data. To succeed, a data management program requires a partnership between the business and technology groups. Given the broad focus of data management, an effective program relies on a combination of people, process, and technology.

Enterprise Data Management Maturity Model

The Capability Maturity Model for Software (also known as the CMM and SW-CMM), published by the Software Engineering Institute (SEI) and Carnegie Mellon University, is a well-established model that defines software development maturity of organizations based on procedures and processes. However, it does not address the maturity of an organization with respect to how data is managed. Larry English adapted the CMM to data quality in his book titled, *Improving Data.*

This new maturity model coined as the "Enterprise Data Management Maturity Model" helps organizations identify and quantify their respective levels of data maturity. By assessing an organization's data management maturity, the risks associated with undervalued data management practices of the organization can be assessed. The maturity model also helps organizations understand the benefits and costs associated with a move to the next stage. By implementing change as per an organization's own plan, within a reasonable timeframe organizations can accurately set goals for data maturity.

Understanding the maturity model can help organizations control the evolution from stage to stage. However, organizations need to know at what stage they currently are operating, allowing them to understand how and when to move to the next stage. The stages of the Enterprise Data Management Maturity Model are:

Stage 1—Unaware
Stage 2—Reactive
Stage 3—Proactive
Stage 4—Predictive

Once assessing the current level is done, organizations need to determine what stage is appropriate for their organizations and establish the actions and priorities for improvement.

The Enterprise Data Management Maturity Model has four stages of data management maturity in a continuum (see Figure 3.9). The movement from one level to the next is a continuous process and will not happen all at once. Often, different parts of an organization may be at different stages of the maturity model. And ultimately, an organization or parts of an organization may not choose to move to a higher maturity level if the costs of the move outweigh the benefits. The potential rewards escalate as enterprises move from level to level. And, each stage of the model

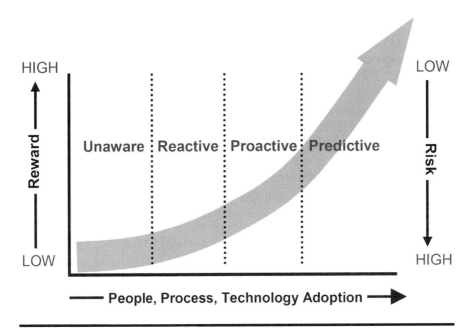

Figure 3.9   The Enterprise Data Management Maturity Model

requires certain contributions and investments. An organization's stage of development depends on many factors including:

- *People:* Who is involved and what contributions they made at the current stage, and what they must contribute to moving forward.
- *Process:* What activities are performed at the current stage, and what must be performed to move forward.
- *Technology:* What investments in technology were made at current stage, and what must be made to move forward.
- *Risk and reward:* What risks does the organization face at the current stage, and what could it gain from moving forward.

For enterprises who have achieved significant results from improved data, data maturity is not just a technological approach to understanding and correcting data. It is also about implementing a sound process to collect and manage information over time. The Enterprise Data Management Maturity Model recognizes that an examination of people, processes, and technology identifies ways to improve data integrity over time.

Appendix B examines each of these stages, and the characteristics of organizations within each category. And for each of the first three stages, there are suggestions for progressing within the model, from "Unaware" to "Predictive."

Access Data

It is important for an organization to recognize that a good understanding of its customers is useful only to the extent to which this knowledge can be translated into real business practices. Business intelligence refers not only to the data analysis in itself, but also to how the results from the data analysis relate to everyday business decisions and how the recommended actions stemming from the analysis translate into live actions.

It is therefore important for functional departments to interact with the data analysts constantly throughout the process. That way, when the data analysis is complete, the functional personnel will already be in tune with the issues the organization is facing, and will be able to develop plans to capitalize on opportunities and strategies to mend weaknesses quickly and effectively.

Dating back to the days of mainframes, minicomputers, and COBOL reporting, business decision makers relied heavily on the Information Technology (IT) staff to help them understand day-to-day business workings via the creation and delivery of standard and custom reports. As technology evolved, more data became available in transactional systems and in turn the need to access and understand this information grew.

This marked the beginning of the never-ending reporting backlog; the IT team created standard predefined reports by pulling data from the transactional systems, but the standard reports rarely provided the necessary information. Business users constantly demanded more data and additional one-off custom reports. The resulting backlog meant business users often had to wait days or even weeks for reports. Often these custom reports did not provide the appropriate information but would prompt another series of requests for additional custom reports, starting a vicious cycle, placing IT in a difficult position, to say the least.

Data access and the ability to support scheduled and ad hoc reports from both transactional and reporting databases have serious implications for data architecture and management. Data access is fairly straightforward but its importance cannot be overlooked. Without complete access, the information consumers (both end users and applications) will get an erroneous picture. For years, conventional business intelligence and data warehousing wisdom held that business intelligence applications should access information exclusively from a data warehouse to avoid putting a drain on operational systems.

This, combined with the fact that most BI tools were optimized to report from a data warehouse and could not report directly from operational systems, has meant that enterprises have undertaken massive data-warehousing projects at a huge price and considerable risk, in order to have a suitable data infrastructure for business intelligence and reporting. Even with ETL tools to automate the movement of the data to the appropriate warehouse, operational data store, or data mart, project cycles and solution paybacks have proven lengthy.

The problem with this, in addition to its huge cost, is the amount of latency it creates. There are times when real-time does mean instantaneous, especially when someone is trying to pinpoint a problem. Even if the data warehouse and application reports are updated on the hour, it is not good enough. As business cycles continue to shorten, and more businesses adopt real-time models, these situations will become more frequent. Without the ability to access live data, it is impossible to know whether you have the right answer. This could be the difference between making earnings estimates and receiving a warning.

Analytics Moves to Desktop

During the past decade, the sharply increasing popularity of the personal computer on business desktops has introduced many new options and compelling opportunities for business analysis. As personal computer technology became more widely available to the business community, information usage patterns changed, triggering a transformation in BI thinking.

Executives, managers, and business users throughout the organization faced a need to make decisions more quickly to survive. Thus, individual users, and even whole departments, started to take control of their own data requirements by using spreadsheets as a central repository for information. This caused the BI model to evolve in a manner to incorporate reports, spreadsheets, and analysis tools to understand the new situation.

For a while, organizations were satisfied with ad hoc reporting tools that simplified the reporting process, enabling nontechnical business users to create their own reports and reducing the burden on IT staff. These tools, however, did not completely insulate the end users from technical intricacies; although graphical in nature and relatively easy to use, they gave them direct unregulated access to a very complex environment. Few end users were able to efficiently and correctly satisfy their own reporting requirements, and IT staff found it difficult to maintain control over the environment.

Given today's economic climate, with more organizations struggling to get by with less, adding additional resources dedicated to analysis is not a reasonable option. However, making current analysts more effective is an avenue worth pursuing. This can be achieved by investigating platforms and analytic environments that leverage advanced techniques capable of rapidly scanning through the various combinations and revealing interesting anomalies and patterns that can be further explored. Until recently, solutions supporting these requirements have been evolving slowly.

Various estimates and surveys suggest that most organizations have provided business intelligence applications to only 5 to 7 percent of the people who could use them. In this small number, two-thirds are casual users at the bottom of the usage pyramid and really only need to see information in a report, whereas the business users in the middle, having a bit more technical skill and needing their information in a form that can be easily manipulated, constitute 25 percent. Still, only about 5 to 7 percent of users possess the highly technical competency to fully utilize this information format, thus pointing to the need for an analysis tool. Even today, as our massive amount of information continues to grow, most data is still being analyzed in spreadsheets.

Many reporting and analytical packages provide the capability to graph and chart measures against particular views. For example, those interested in operations would like to see the profitability of a particular asset across different geographies. After reviewing the results, the end user most likely will ask another question and subsequently review those results. The end user formulates ideas and opinions based on this type of interaction. With a manageable number of possibilities, this type of session is likely to identify actionable business insight.

Smart business decisions are the direct result of having the right information at the right time. However, in most environments decision makers are only provided predefined standard and parameterized reports, or in some cases ad hoc query and reporting tools to make "on the fly" reports. However, complexity in the data schemas, nuances in data encoding, calculation, and the time it takes to learn more advanced decision support tools often deter most decision makers from using these tools, leaving users with predefined reports and reliance on an intermediary to access deeper information.

During the past decade, the sharply increasing popularity of the personal computer on business desktops has introduced many new options and compelling opportunities for business analysis. The gap between the programmer and end user has started to close as business analysts now have at their fingertips many of the tools required to gain proficiency in the use of spreadsheets for analysis and graphic representation. Advanced users will frequently use desktop database programs that allow them to store and work with the information extracted from the legacy sources.

Today, more and more sophisticated tools exist on the desktop for manipulating, analyzing, and presenting data. Many desktop reporting and analysis tools are increasingly targeted toward end users and have gained considerable popularity on the desktop. However, there are significant problems in making the raw data contained in operational systems available easily and seamlessly to end-user tools. To address business problems in real-time, organizations must deliver more information to more people, both inside and outside the enterprise, including employees, managers, partners, suppliers, and customers.

The downside of this model for is that it leaves the data fragmented and oriented towards very specific needs. Each individual user has obtained only the information that he or she requires. Not being standardized, the extracts are unable to address the requirements of multiple users and uses.

One way of solving this problem is to find a common data language that can be used throughout the enterprise. Still, this approach to data management assumes the end user has the time to expend on managing the data in the spreadsheets, files, and databases. Many of these users may be proficient at data management, however, most undertake these tasks as a necessity. And given the choice, most users would find it more efficient to focus on the actual analysis.

DATA USAGE

Decision makers are literally buried in reports of all shapes and sizes, but never feel they have exactly what they need. This situation is epitomized by the phrase, "data rich but knowledge poor." In any given organization there are many different decision makers using a variety of different methods to review and analyze data.

Reporting, ad hoc or predefined, and analytics all have their place in the decision-making process depending on the specific needs of the user at any point in time. Some groups (e.g., clerical) need simple information delivery in predefined reports. Other groups, specifically those with decision-making responsibility, sometimes need basic information delivered to them, whereas on other occasions they need a much deeper understanding to make the proper decisions.

It is critical to understand exactly how the target user(s) interact with the information prior to determining a delivery mechanism. Below are samples of data usage and analysis ranging from simple to more complex. Simple forms of data usage include:

- *Reading:* Simply looking at a predefined report for specific information and basic facts pertaining to the subject. Reporting servers are designed to facilitate this need. For example: the inventory report is retrieved and the user quickly scans through certain products looking for quantity on hand.
- *Monitoring:* "Reading" but in a monitoring manner, meaning information is looked at every day or every five minutes, and so on. Real-time reporting and alerts are great technology for this need.

However, as analysis grows more complex, simply delivering the information in a predefined report grows inadequate:

- *Trend analysis:* Putting new information in context with historical data to understand growth or change in a time series. This is needed to better understand where the numbers are going and to determine if this is a trend or an exception.
- *Root cause analysis:* Ad hoc analysis to determine what factors are driving a specific number. Decision-making responsibility means deciding when things need to be changed and specifically what to change. This requires knowledge of more than just the high-level issues; understanding of the underlying driving factors is critical.
- *Statistical analysis:* Correlation, regression, affinity, data mining, and so on. These analysis techniques are used to more deeply understand information and uncover hidden insights in historical data. Performing this type of analysis requires skilled training and thus, the tools are not widely used. However, the insights revealed can be extremely valuable to decision makers.
- *Predictive analysis:* Extrapolation, risk analysis, decision modeling, what-if scenarios, and the like. This complex analysis is performed in order to drive desired outcomes in the future. Examples include: price modeling, derivative risk exposure, forecasting, and sales pipeline analysis.

Taking the time to understand what decisions will be made and how users will interact with and analyze the information will help immensely in determining the most appropriate delivery method.

Reporting—Designed for Distribution

At its core, reporting technology is designed for information distribution. Whether it is ad hoc or predefined, the reporting paradigm addresses information delivery, publishing, and distribution. That is why reporting products have menu items that predefine the query, format the resulting data, and then publish that report to the people for whom it was designed.

When describing BI technologies, reporting is often lumped together with analysis to categorize a technology as "reporting and analysis." This makes some sense because analysis is what users ultimately do with a report once it is received. If the user knows ahead of time what data is needed and considers the job complete once the report has been published, then reporting technology is the right tool for the job. Every information consumer in the organization, from executives to production workers, will use these reporting systems.

However, if the user's job is to make decisions, this "delivery" model may only be adequate in certain circumstances and be significantly inadequate in others. The most popular feature in any reporting system is the "Export to Excel" button simply because the decision-maker's job with the data is not done when the information is delivered; the job has just begun.

The exact distribution of information users throughout the organization will vary by industry and by each individual organization. For example, the business user distribution in a telecom service provider will look much different from that of a labor-intensive manufacturing organization. In most cases the executive-level decision makers would run in the 3 to 8 percent range. The next two levels, tactical decision makers and production and clerical workers, tend to range widely depending on industry, perhaps 40/55 percent, respectively, on the high end, or down to a 25/70 percent split on the other end.

It is important to recognize that although the largest group of information users may be in the production/clerical role, these people are not involved in decision making. They do have information needs, but given their narrow scope and predefined jobs, standard report systems will meet their requirements. Thus, high ROI decision support systems should focus on the top two groups, executives and managers/knowledge workers, where the decisions are being made.

Operational Reports—Banded Layout

The vast majority of enterprise information dissemination is in the form of traditional operational reports. These time-tested reports display volumes of tabular data organized into a hierarchy of increasingly finer levels of detail. BI tools should be adept at quickly organizing massive amounts of operational data into well-organized tables (see Figure 3.10).

Banded layout is the most familiar report layout technique today because it is the predominant vehicle for operational reporting. It has been the cornerstone of all leading report writers for years. The reason for the long-term popularity for operational reporting is that banded layout can very rapidly organize lots of data into a natural hierarchy of information that users can easily navigate.

The term "banded" refers to specific horizontal bands superimposed on the report layout screen that dictate where data should be automatically summarized for page headers and footers, report headers and footers, and within user-defined groupings or hierarchies that map to the business organization. Grouping and summarizing techniques automatically come with banded layouts.

The banded report layout is truly optimized for traditional operational reporting with its dense hierarchical information presentation, multipage

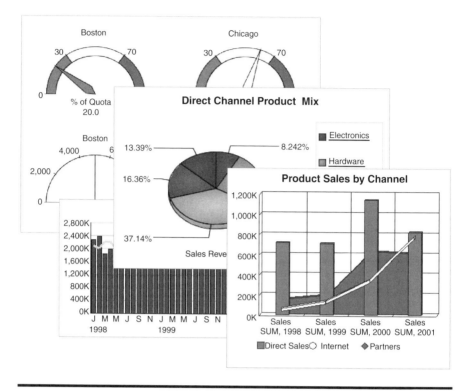

Figure 3.10 BI graphic reports

repeating sections that cover entire ranges of enterprise operations, and bias for fine printing optimization.

Scorecards and Dashboard Reports—Zone-Based Layouts

Scorecards and dashboards are designed to deliver maximum visual impact to the user in a format optimized for quick absorption. BI scorecards combine tables, graphs, gauges, and other graphical indicators, conditional formatting, freeform labels, borders, and background colors to achieve this impact.

Because enterprises need to convey key performance indicators based on a complete view of all financial and operational data, scorecards have to extend far beyond summary level information. Scorecards built on the BI platform utilize the full wealth of enterprise data and take advantage of its scalability, transaction level data access, and ad hoc capabilities.

Zone-based layout, unlike banded layout, is optimized for the creation of scorecards and dashboards. The term "zone" refers to the graphical building block technique that allows users to lay out entire tables and graphs onto a page or screen, where each of these tables or graphs has a presentation behavior of its own. Zone-based layout is optimized for displaying graphical content (as opposed to tabular information in the case of banded layout). Zone-based layout is further optimized for onscreen display, where users have scroll bars to move around the report and desktop publishing quality printing is not a requirement.

Users can build scorecards and dashboards on the fly using simple drag-and-drop techniques to insert multiple reports, graphs, text, hyperlinks, and images onto the report layout screen. These objects may also be arranged anywhere in the layout and automatically adapt to the size and shape of their content, moving other objects dynamically to fit.

Blending Reporting with Analytics

True analytic software is different in that it is designed to facilitate the analysis of information: inspection, exploration, and question/answer probing that coincide with the human process of assimilating data. Although the software will allow decision makers to specify a query, format it, and distribute it, this is not what analysis tools do best. Menu items in analysis software facilitate exploration, root cause and historical context analysis, and use advanced data visualization to help users grasp the relevance of the data.

Reporting and analytics together provide an organization with the perfect combination of data distribution and necessary tools to understand information. These technologies do different jobs and when deployed correctly can work in conjunction to fulfill the spectrum of an organization's information needs.

Unfortunately, organizations commonly utilize one or the other of these technologies to solve problems the software was not specifically designed to solve. This means using the older, more widely understood reporting technology to address users' needs in data understanding. This leads to the all too common issue of "spreadsheet anarchy," spread marts, and an overall dissatisfaction with the organization's decision support efforts. Because analytic software has become available more recently and is less understood, it is rare that this technology is used for information delivery, but it does happen, creating a whole new series of issues.

When these two technologies work seamlessly together, users can get information delivered to them in a variety of ways. When they need to know more about the information, they can simply click on a data point to launch an analysis and quickly determine the driving factors that are critical to good decision making.

With the emerging class of open standards and next generation reporting and analysis offerings, this integrated approach can be achieved with surprisingly affordable solutions. The key is to pick a reporting solution designed to be a world-class reporting system and pick an analysis technology that is equally best in class with offerings that can work seamlessly together.

ENTERPRISE PORTAL (EP)

Enterprise Portals (EPs; Figure 3.11) are applications that enable enterprises to unlock internally and externally stored information, and provide users a single gateway to personalized business information needed to make informed business decisions. Enterprise portals can be seen as a browser-based system providing ubiquitous access to business-related information in the same way that Internet content portals are the gateway to the wealth of content on the Web.

EPs have attempted to address the key needs of a sophisticated BI environment where, at a high level, end-user requirements traditionally go

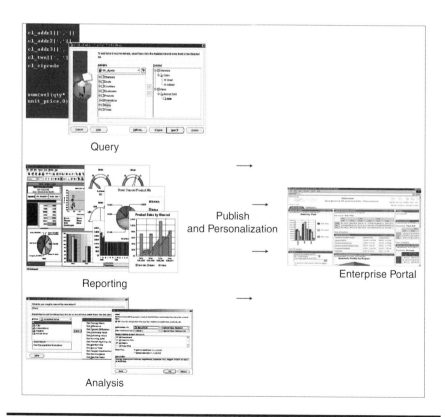

Figure 3.11   Enterprise portal

unmet or almost always fall short. To cater to this demanding condition EP must be capable of delivering seamless user-focused administration, providing a feature-rich user experience, and integrating BI applications delivering key business functionality. Finally, it should do all of this in a timely cost-effective manner bridging the gap between the technology and the user.

What Is Enterprise Portal (EP)?

Enterprise portals can be defined as a Web application that integrates many different corporate and external resources into one scalable user experience, using primarily Internet standards for interapplication communication and integration.

EPs are an emerging business tool, an amalgamation of software applications that consolidate, manage, analyze, and distribute information across and outside an enterprise (including business intelligence, content management, data warehouse and mart, and data management applications). EP applications combine, standardize, index, analyze, and distribute targeted relevant information that end users need to do their day-to-day jobs more efficiently and productively.

Functionalities provided by EP are:

- Robust search across all repositories
- Taxonomy support
- Content management/aggregation
- Personalization
- Application integration/development
- Web services

Why EP?

EP systems provide organizations with a competitive advantage. Top management is just realizing the competitive potential lying dormant in the information stored in its enterprise systems. The key to unlocking it and combining it with information from external sources is the EP tool. The benefits include lowered costs, increased sales, and better deployment of resources.

EP systems provide enterprises with a high Return on Investment (ROI). EP products help enterprises cut costs and generate revenues. Organizations become more proactive, agile, and competitive in the marketplace because they have an integrated application that provides accurate timely information for better performance analysis, market segmentation and targeting, forecasting, customer relationship management, supplier relationship management, workflow management, and so on.

EP systems provide access to all. The Internet provides the crucial inexpensive and reliable distribution channel that enables enterprises to make the power of information available to all users (employees, customers, and suppliers). Distribution channels include the Internet, intranet, and broadcasting. A critical element to the success of EP applications in corporations is the accessibility of information to all users (employees, customers, and suppliers). Organizations will need to use both "publish" (pull) and "subscribe" (push) media to ensure the right information is available or distributed to the right people at the right time.

Effectively Integrating EP

The key challenge of effectively integrating an EP may be broken into several factors.

- *Administration:* Administration is a challenge because, currently, this is a centralized process whereby an umbrella organization administers the whole.
- *Application integration:* Application integration is demanding because the portal must keep up with new releases and potentially a low-level functionality inherent in most applications.
- *BI features:* Features may or may not exist in some EPs that are critical to specific components of a business.
- *Cost and schedule:* Cost and schedule are probably the most challenging components of an EP deployment. This is due, in large part, not to the actual physical deployment but more from a source-selection, planning, and infrastructure standpoint.

BI Features

The reality of an enterprise environment dictates that BI be delivered with a user focus and not be segmented by technology. EP can bridge this gap of tools to deliver BI-derived information at a consolidated level by providing the ability to integrate across various BI tools. Some of the key requirements from EP, needed to achieve this goal, include:

- BI product-independent KPI alerting
- BI product-independent dashboard customization
- Consistent report administration across BI products
- Security management across BI products
- BI product-independent user/group management
- BI product-independent usage tracking
- BI application integration

4

BUSINESS INTELLIGENCE: TYPES

Ultimately, Business Intelligence (BI) is about providing information to as many users as possible, to help them cut costs, increase productivity, or drive more revenue. The origins of BI are from the data warehousing, reporting, and query tools and applications that came bundled with enterprise applications and the tools and applications that arose to begin integrating all this data.

BI applications have evolved dramatically in the past decade as organizations' access to, and appetite for, information grew exponentially. Today BI continues to grow at an even more rapid pace as more and more organizations begin to understand the importance of data in delivering a competitive edge.

The dramatic expansion of data warehousing combined with the widespread adoption of enterprise applications, such as Enterprise Resource Planning (ERP) and Customer Relationship Management (CRM), and the overall increase in computer literacy, fueled this exponential demand for BI reporting and analysis applications.

BI applications, which were rudimentary in their early years of evolution, have progressed a lot, from operational basic reports generated by mainframes, to statistical modeling of promotional campaigns, to multidimensional OLAP environments for analysts, and to dashboards and scorecards for executives. Also with this growth comes the demand for more ways to report on and analyze data.

During its formative period, organizations actively discovered many new ways to use their data assets for decision support, operational reporting, and process optimization. And technologists reacted by building niche software to implement each new pattern of application that organizations demanded. These patterns of applications resulted in products centered exclusively on one type of BI or another.

Most large enterprises have purchased many different BI toolsets from many different vendors, with each tool targeted at a new BI application and each tool delivering user functionality focused on only one of the types of BI.

The most sophisticated and interactive tools of BI are used by relatively small groups of users consisting of information analysts and power users, for whom data and analysis are their primary jobs. Less interactive BI tools deliver basic data and results that are applicable to very large user populations ranging from senior executives all the way to staff personnel.

It is generally accepted now that from the top management to the support staff, every organization employee analyzes business data to some degree, in some fashion. Their analyses may be deliberate and exploratory, they may be triggered automatically by threshold conditions, or they may even be so embedded in everyday systems that their existence as BI per se may not even be recognized. One thing is clear: successful organizations make maximum use of their data assets through BI technology.

MULTIPLICITY OF BI TOOLS

Leading organizations have recognized the benefits of putting information into the hands of all their employees, regardless of job title or function. By nature, organizations are comprised of a variety of "communities," each of which needs different kinds of tools to do its job. When it comes to business intelligence, analysts use OnLine Analytic Processing (OLAP) and other instruments, specialists use data mining and predictive modeling, and nearly everyone else relies on reporting solutions. And spreadsheets are available to all.

What's more, an organization's various operational entities—divisions, departments, and business units—typically have long depended on their own private stores of data and information, tailored to their needs, and not shared. Today, most large organizations have more than a dozen different BI technologies installed and in use somewhere in their organizations.

Reasons for this gross proliferation of diverse technologies are limited capabilities of the majority of deployed BI tools and the practice of empowering individual departments to choose whatever BI technology they wanted. BI technology vendors reacted to this purchasing behavior by developing departmental BI products. These tools which support just a single type of BI at a time have limited user scalability, limited data scalability, and operational characteristics more attuned to casual small-scale operation. These tools could only support unit or departmental scale applications, which led to different BI applications being deployed in isolation and led to multiplicity of resources.

The end result of departmental BI purchasing has been the massive proliferation of islands of BI that is seen in most organizations. As in the past where IT resources for e-mail and client/server computing were fragmented, departmental aspirations gave way to enterprise aspirations, and the departmental technology gave way to enterprise technology. The same is happening in the case of BI: in technology-forward enterprises, departmental BI tools are giving way to the enterprise BI platform.

The Problem with Multiple BI Tools

The trouble with such balkanization is obvious: way too many resources wasted on administrative overhead, poor data quality, difficulties collaborating, and decisions made with inadequate information. There are five key issues that make isolated departmental islands of BI and the use of disparate departmental BI tools a great drag on an enterprise.

Scalability—Departmental BI Lacks User and Data Scalability

Based on the universal success of these BI applications, enterprises are now emboldened to take it to the next level and are refocusing their aspirations from department-level plans to enterprisewide strategy, in favor of achieving greater impact with BI applications. And that means delivering much richer reports and analysis, from much larger pools of data and delivered to many more users. Unfortunately, most departmental BI tools today cannot scale to these new levels. The very nature of their underlying architecture prohibits them from analyzing terabytes of data and delivering or expanding.

Inconsistent Versions of the Truth

Multiple islands of BI result in multiple inconsistent metadata repositories. Multiple independent islands of departmental BI applications work fine when the number of applications is small. When there are few applications, there is little overlap in analytical or reporting domains; inconsistencies in data definitions, and in metric usage, and business models are not readily evident. However, when the number of BI applications achieves a certain critical mass within the enterprise, there is an inevitable overlap in analytical and reporting domains. It becomes inevitable that multiple reports from multiple independent BI applications present inconsistent information, preventing a consistent version of the truth. As the number of applications increases, these inconsistencies undermine the integrity of all the BI applications.

Users Being Forced to Use Multiple BI Tools

Multiple user interfaces are problematic. When the number of BI applications is few, any given person only uses one of those applications, and hence only uses the one BI technology associated with that application. As the number of BI applications increases, more and more people will be accessing multiple BI applications, and hence using multiple different BI user interfaces to view reports and manipulate those reports. This means that BI users need to learn different ways to do everything, including such common actions as finding reports, running reports, scheduling reports, editing reports, saving reports, sharing reports, answering prompts, sorting the data, and pivoting the data. Nonetheless, this is the situation in which most enterprises are now finding themselves.

Need for Richer User Experience Encompassing Multiple Departments

Single-purpose departmental BI tools are unable to mix and match information and processes from multiple departments. Most departmental BI applications are not able to work outside the boundaries of their department. However, there is really a natural user flow starting with information presentation, to problem isolation, to full investigation, to advanced analysis, to proactive tracking and alerting; only technical boundaries that isolate the departments are purely artifacts of single-purpose BI technology, and have nothing to do with user preferences or natural application boundaries. With more and more enterprises looking to leverage BI on the enterprise, the widescale user will need more functionality and richer user experience encompassing multiple departments.

Excessive Cost of Managing Multiple BI Technologies

Finally, managing and supporting many diverse BI technologies is a costly proposition. IT organizations suffer excessive redundant costs not only in handling BI tools, but also in resolving technical and operational intricacies these tools bring with them.

Furthermore, with multiple BI tools, enterprises need to train people in the development and operational aspects of each BI technology, establish technical support teams to specialize in each capability, manage contracts with each BI vendor, coordinate new version upgrades with their versions of database software, server operating systems, workstation operating systems, browsers, Web servers, and firewalls.

And that's not all: enterprises must manually synchronize metadata overlaps between multiple BI technologies, such as security, business definitions, metric definitions, and user profiles. For forward-looking

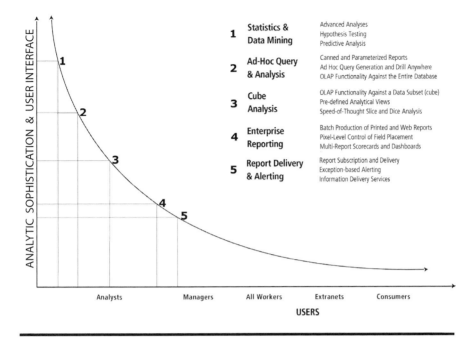

Figure 4.1   Types of BI

organizations, the cost can be even larger as they may like to create knowledge base and higher competencies that will be a huge cost if multiple technologies are supported.

TYPES OF BI

BI types (shown in Figure 4.1) include enterprise reporting, cube analysis, ad hoc query and analysis, statistical analysis, and data mining, alerting, and report delivery.

Enterprise Reporting

When an enterprise wishes to distribute standard operational reports or financial reports to all stakeholders in the organization, enterprise reporting is used. It includes report formats for operational reporting and scorecards and dashboards targeted at information consumers and executives.

Organizations have found clear returns on their investment in operational and financial reporting. Enterprise reporting is the most widespread type of BI used since the 1950s, evolving from its earliest adoption as mainframe reports to today's Web-based reports.

Cube Analysis

Cube analysis is the type of BI ideal for basic analysis that can be anticipated in advance. It is OLAP slice-and-dice analysis of limited data sets, targeted at managers and others who need a safe and simple environment for basic data exploration within a limited range of data. For example, the analysis of sales by region for certain time periods, and the analysis of sales by product and by salesperson, for instance, could be useful to store managers looking for some underlying details on performance.

Ad Hoc Query and Analysis

Ad hoc query and analysis is the type of BI that enables true investigative analysis of enterprise data, down to the transaction level of detail. It is full investigative query into all data, as well as automated slice-and-dice OLAP analysis of the entire database, down to the transaction level of detail if necessary. It can serve as a vital tool for information explorers and power users.

Statistical Analysis and Data Mining

Statistics and data mining is the type of BI used to uncover subtle relationships (e.g., price elasticity) and forecast projections (e.g., sales trends), using set theory techniques, statistical treatment, and other advanced mathematical functions. It is full mathematical, financial, and statistical treatment of data for purposes of correlation analysis, trend analysis, financial analysis, and projections.

Alerting and Report Delivery

A report delivery and alerting engine allows enterprises to distribute vast numbers of reports or messages on a proactive and centralized basis, as well as allowing users to self-subscribe to report distributions. It is proactive report delivery to and alerting of very large populations based on schedules or event triggers in the database. Report distribution can be initiated on a scheduled basis, as well as on an event-triggered basis, such as a metric's value falling below a target threshold.

MODERN BI

A modern BI platform is going to tie into any corporate database platform, and will have a Web-based user reporting and querying interface as the front end. A critical component of a good BI plan is that the end-user tool must be easy to use. There can be a typically browser-based model which users will already understand as the basic user interface.

BI should take the job of analysis from distant analysts and put it into the hands of as many decision makers in an organization as possible. A manager or sales rep should not have to ask an analyst to run a report or create a new report with BI. Instead the managers or reps should have an easy-to-use tool they can master using skills they already have to run their own reports and create ad hoc queries on their own. The common thread here is escalating complexity, and business intelligence tools are being adapted to help both business and casual users—often consumers of information rather than producers of analytics—cope with it. Compare this to OLAP, for example, and it would not be reasonable to expect an end user to want to use it.

Thus, BI applications are getting key enhancements such as Web-enabled information delivery and dashboards so that non-analysts can extract intelligence:

- To improve competitive response via use of better metrics and controls, which can uncover hidden costs, and by increasing the velocity of decision making
- To comply with corporate governance regulations including SOX, the Sarbanes–Oxley Act, that force organizations to be able to access detailed current business information to an unprecedented degree
- To upgrade existing ERP implementations so that managerial processes are made more effective

For most BI applications other core requirements will be real-time analysis, visualization features, forecasting, and decision support. It will probably also need to tie into CRM application and production systems as well (to analyze and predict customer behavior and to look at production costs and efficiency). If an organization has a Six Sigma program or similar quality management philosophy, expect a BI application to need to interact with that tool too. As can be seen, there is very little that a BI plan might not encompass.

Indeed, BI's role cannot be overstated, especially as it is applied to enterprisewide issues. With the enterprisewide integrated BI offerings organizations can:

- Get more out of ERP, CRM, and supply chain management systems.
- Eliminate or repurpose redundant BI solutions.
- Standardize on fewer BI platforms, tools, and applications to satisfy both the analytic reporting needs of power users and the enterprise reporting needs of casual users.

Table 4.1 Enterprise Business Intelligence Platform

Unified User Interface
- No disorientation caused by multiple UIs
- Only one BI paradigm for users to learn

Unified Five Styles of BI
- BI applications can exhibit any or all types of BI
- All applications behave the same way

Unified Services Backplane
- Common metadata
- Common security
- Common prompting
- Common scheduling
- Common privilege control
- Common administration
- Common governing

Enterprises today need a BI technology that can support any or all of the types of BI for any given BI application and are increasingly dissatisfied with functional limitations of BI tools (see Table 4.1). One of the reasons behind the slow takeoff of BI in the past, even though it provides exceptional benefits, is that most available BI technologies were fragmented along types of BI.

Fragmented BI technologies perpetuate a multiplicity of BI applications, different tools and applications for different users and functions, while designing an all-inclusive BI solution. This deployment of multiple tools within the enterprise imposes different user interfaces, different metadata, and different security for each type of BI. This multiplicity leads to an increase in costs of ownership, decline in user acceptance, and decrease in richness of each BI application while it escalates the need for IT resources for operation and maintenance.

Enterprises need architecture that can deliver all five types of BI. An architecture that can deliver all five types of BI must have three key features.

1. Can be mixed and matched seamlessly for users, where the addition of each new BI type adds functionality to the users' existing reports
2. Can be expressed through a single unified user interface to maximize ease of use and user acceptance
3. Can be delivered on top of a single integrated platform that unifies the metadata, security, and user profiles, ensuring a single version of the latest information throughout the enterprise and thus minimizing administration and maintenance efforts by IT

The next sections examine these types of BI in detail.

Enterprise Reporting

Enterprise reporting, in essence, provides business intelligence to the masses. As a result, it is the most prevalent type of BI, encompassing a vast array of operational reporting directly from ERP and CRM systems, as well as scorecards and dashboards of overall business performance.

Enterprise reporting must be designed for information consumers including those individuals at all organizational levels and across all job functions in the enterprise and also including supply chain partners and even customers. These individuals get their reports by accessing them on demand through their Web browsers (Web-based reporting), as well as by receiving distributions that are pushed to them via e-mail or print delivery.

The single most dominant characteristic of any enterprise reporting system is its ability to produce highly flexible report formats, so that data can be presented in whatever form is most consumable to a wide range of information consumers.

Other than the dominant characteristic mentioned above, the capabilities of enterprise reporting technologies revolve around the following key areas.

Support for Different Forms and Types

Report-writing products typically either can deliver operational reports well or can deliver scorecards and dashboards well, but for an enterprise BI tool, both play an important role and cannot be compromised.

An enterprise BI is expected to deliver from scorecards and dashboards at one extreme, all the way to operational reports at the other extreme, and the many variations in between. Furthermore, it must be pixel perfect (can be delivered onscreen) as well as print perfect (can be delivered on paper) not requiring any programming.

BI with advanced enterprise-level architecture must be designed to deliver both operational reports and scorecards and dashboards easily from a single platform. In fact, BI should allow users easily to develop the common forms of enterprise reports that range from highly graphical scorecards and dashboards for executives to densely populated tabular operational reports for all personnel. In between these extremes are classic business reports for business unit managers, managed metrics reports for business unit leaders, and invoices and statements for customers and partners.

Typical report-writing products are optimized for either onscreen presentation or for print generation, but not both. Moreover, many of these report writers require extensive programming to deliver all but the most rudimentary reports. However, enterprise-level BI must be designed to be able to cater to demand by making available desktop publishing quality reports.

Support for Personalization and Customization

Another feature that must form part of the modern enterprise-level BI tool is making enterprise reports more compelling, personalized, and relevant for each individual.

The goal of every enterprise reporting project is to inspire people to use information in their day-to-day work activities. This is possible if enterprise reporting design can incorporate four personalization levers.

Parameterized Reporting — User-Defined Report Contents

Parameter-driven reports allow the user to answer a number of questions prior to running a report and those answers dictate what content will be displayed in the report. It allows users to generate just the data they are interested in seeing at that moment. Users get the ability to tailor:

- Report contents—by picking filters to be used, identifying metric limits and thresholds, and selecting benchmark values against which to compare reported numbers
- Report layout—by controlling the arrangement of columns, picking variables to act as cross-tab organizers, and choosing the page by grouping of reports
- Workflow—by selecting and defining triggers for alerts, and directly importing results from other reports

Automatically Customized Content — Allowing One Report to Serve Thousands of Users Automatically

A common challenge in enterprise reporting is how to deliver similar information to very large populations of users economically, where each user needs to see a different slice of the data.

With BI tools capable of automating content customization, IT administrators only need to create one master report and the BI architecture automatically creates personalized variations of any report definition based on the group affiliation and role of every report user. Hence, any report can be sliced along many dimensions automatically to make the report content immediately relevant to the recipient users.

Personalized User Interface — Matching Interface Functionality with User Skill Level

Enterprises face a challenge when striking a balance between exposing rich functionality to power users, while at the same time giving novice users a simple enterprise reporting environment that will not overwhelm them.

With personalization of the user interface this problem can be resolved. User profiles automatically adjust the Web interface to accommodate users with different skill levels. User profiles determine exactly what functionality will be exposed to each user or user group. So it is easy to give report designers a user profile with maximum functionality, while the report consumers receive a user profile with just enough functionality to do their jobs easily.

Support for Wide Reach, High Throughput, and Access across All Touch Points

Information consumers are everywhere, in executive offices, in office cubicles, on warehouse floors, at loading docks, at customer sites, at suppliers' offices, and at customers' homes. Effective enterprise reporting systems must be able to reach all of these users, wherever they are and with sufficient power to generate tens of thousands of reports per hour if needed. This can be done by reaching all users through the touch points of their choice and making reporting ubiquitous and convenient throughout the enterprise. Reports must be accessible through every touch point including Web browsers, networked printers, e-mail, networked file servers, and corporate portals.

Cube Analysis

Cube analysis delivers the simplest form of analysis, allowing anybody to analyze data. Cube analysis is used most often by people such as managers who have a deep interest in understanding the root causes underlying the data in reports, but do not possess the skills for full ad hoc investigation of the databases.

Cube analysis lets people flip through a series of report views, using the now standard OLAP features of page by, pivot, sort, filter, and drill up/down. These OLAP features, which were first introduced in the early 1990s, allow users to slice and dice a cube of data, or analysis cube, using simple mouse clicks.

The term "cube" refers to a subset of highly interrelated data that is preorganized to allow users to combine any attributes in the cube (e.g., stores, products, customers, suppliers) with any metrics in the cube (e.g., sales, profit, units, age) to create various two-dimensional views, or slices, that can be displayed on a computer screen.

To implement cube analysis functionality, most OLAP vendors use custom-made proprietary cube databases. This technique is known as Multidimensional OLAP or MOLAP. Unfortunately, the cube databases have very small data capacities—less than 0.01 percent of real relational databases—because they do not have the technical underpinnings of real relational databases. Nonetheless, this capacity limitation was not initially perceived as a problem because most early departmental BI applications only needed between 10 Mb and 100 Mb of detailed and summary data. Problems due to limited cube data capacities began occurring when enterprises found they needed to deploy hundreds of overlapping cube databases to cover all the combinations of data subsets, summarization levels, and security privileges for different user groups across multiple applications. These ever-growing collections of cubes have become known as "cube farms." Cube farms create an immense burden on the IT groups that have to generate the cubes, precalculate the summarizations, distribute them to users, and retire them when their data becomes outdated.

However, OLAP functionality can be implemented by modeling the relational database as a "virtual multidimensional cube" with a technique known as Relational OLAP or ROLAP. BI users have the same OLAP functionality of page by, pivot, sort, filter, and drill, but can do so against the entire relational database. With ROLAP, the data is always the very latest data: there is no limitation of what data can be analyzed, and all users and security work uniformly against the database. ROLAP architecture provides the following capabilities.

High-speed report analysis and manipulations: Analysis of cubes at high-speed performance and powerful slice-and-dice capabilities.

Seamless ad hoc drilling: Seamless drilling capabilities outside the cube domain to anywhere in the data warehouse from summary data to transactional details.

Cube sharing: Transparent and secure sharing across the enterprise of cubes with personalized views.

Automatic creation and synchronization of cubes: Creating cubes on the fly with automatic refresh of data for real-time analysis.

Ad hoc query and analysis: Ad hoc query and analysis is for information explorers and power users who need full investigative power against all enterprise data. These users require the ability to see any possible combination of data. Ad hoc query and analysis—the ability to ask questions about the data underlying a standard report—is an important capability that allows users to explore their data without administrative intervention. In an ad hoc query and analysis environment, business users can generate new or modified reports with control over content, layout, and calculations.

If it were feasible to predesign reports that covered every possible combination of data, then ad hoc query and analysis would not be needed. Practically speaking, this is impossible. Predefining reports with all possible permutations would require the design of tens of thousands and even millions of reports depending on the extent of the database. It would also require the addition of hundreds or thousands of new reports each time a new attribute was added to the database.

The most basic way to support ad hoc query and analysis is to give users the ability to create brand new reports, allowing them to assemble any possible combination of data into a report. In that way, enterprises do not have to predesign all possible report combinations. This is the basic definition of ad hoc query and analysis, and it is the method employed by most BI vendors. Some features of robust ad hoc query and analysis tools are:

Parameter-Driven Reporting and Guided Analysis: Allows users to create radically different reports simply by providing some parameters.

Drill Anywhere: Allows users to surf to any place in the database by using OLAP functionality and following the business model of the data warehouse.

OLAP Analysis Against the Entire Database: Allow users to conduct report manipulations across any part of the database.

Sophisticated Filtering with Power of Sets: Allows users to segment data according to different business criteria in order to refine the data set.

Data Grouping: Allows users to refine the business model without causing any changes to the database or the overall business model.

Statistical Analysis and Data Mining

Statistical analysis and data mining is targeted at professional information analysts, individuals who regularly perform correlation analysis, trend analysis, and projections. This advanced type of BI is achieved by applying mathematical, financial, and statistical functions against enterprise data.

The business insight derived from statistical analysis and data mining is critical for every enterprise. However, specialized data-mining tools are difficult to use: only statisticians with technical training are able to use them. Instead, BI technology was designed specifically to deliver much of the common functionality of data-mining tools, and to deliver it in a way that is familiar and consistent with everyday business intelligence usage.

BI's statistical analysis and data-mining features include the following key areas:

Applying Statistics and Data Mining against the Entire Database: Allows users to pick from more than 200 mathematical, OLAP, financial, and statistical functions to apply against the entire database for more accurate results.

Plug and Play Architecture for Custom Analytical Functions: Allows enterprises to extend the analysis with calculations that are customized to its business.

Seamless Integration with Data Mining Tools: Allows the melding of mainstream BI with the more arcane science of data mining

Sophisticated Collaboration Technology: Leverages analytic functions of relational database management system when present and otherwise supplements with analytic functions of BI analytic engine.

Multi-Pass SQL: Lets users perform any type of analysis—from pricing optimization to rankings and contributions—not possible with single pass SQL.

Alerting and Report Delivery: Designed to proactively distribute large numbers of reports and alerts to potentially very large populations of information consumers, both internal and external to the enterprise. With such a broad mandate, report delivery and alerting must be highly flexible and functional. Most BI vendors offer products that support a minimal form of report delivery and alerting. However, given the need of today's executives more features must be strongly supported.

Report Distribution through Any Touch Point: E-mail, printers, file servers, and portals using HTML, Excel, RTF, PDF, and ZIP containers to carry the report content.

Self-Subscription and Administrator-Based Distribution: Allowing enterprises to mix centrally mandated report distributions with individually driven information needs.

Delivery on Demand, on Schedule, or on Event: Triggering the report distribution through whatever means are most appropriate to the user.

Automatic Content Personalization: Making the report more relevant and secure for all users.

THE ENTERPRISE BI

The ideal enterprise BI solution capable of catering to demand of all types of BI as well as modern-day enterprises is the one that deploys a BI architecture that offers a single unified user interface for all five types of BI, and a single unified services-oriented backplane for all five types of BI.

Single Unified User Interface

The user interfaces for BI are very rich nowadays. They present users with far more than just buttons to run a report. Instead they embody an entire paradigm for interacting with enterprise data. At a minimum, a competent BI user interface establishes a paradigm for:

■ Report Catalogs—Finding and organizing reports
■ Report Creation—Finding report components and laying them out

- Report Execution—Running reports both synchronously and asynchronously, where requests are queued in sequence while the individual waits, or processed in the background while the individual works on other things
- Report Searching—Finding reports or objects to place on a report
- Report Sharing—Sending reports to people on a periodic or one time basis
- Information Analysis—Conducting OLAP functions such as page by, pivot, sort, filter, and drill
- Report Exporting—Exporting the results to Excel, PDF, or flat files
- Report Scheduling—Setting up reports to be pre-run during off hours so they are instantly ready for each user during the day
- Report Printing—Full page layout and printer controls

Single Unified Backplane

The BI architecture should support all five types of BI on a single integrated and unified services-oriented backplane. The idea of a common backplane, offering services and integration for independent engines is not new.

The term "backplane" originates in electronics, where backplanes are circuit boards that contain circuitry, sockets, and a communication bus into which additional electronic boards can be plugged to create a new fully synchronized product.

The hallmarks of this unified backplane architecture are:

Mix and Match Types of BI: So that each new type of BI that is added to the system automatically adds functionality and value to all previously installed types of BI, and all previously built reports and applications

Integrated Services: To achieve great efficiencies and consistency in report design, report maintenance, user administration, and security administration across all five types of BI through use of one unified metadata within the services-oriented architecture

Server-Centric Architecture: To achieve high scale and consistently high performance with great resource efficiencies

In a BI system, the backplane serves as the common platform into which enterprises can plug new types of BI. This way, enterprises can deploy only the types of BI they need, when they need them; and those enterprises can expand with additional types of BI over time as the applications grow and users' skills are ready for them. Such BI architecture is future proof. Enterprises can start small with limited functionality and limited scale, but can grow to include all BI functionality, with the highest scalability, highest performance, and best reliability.

Vision of a Critical BI System

In the current information economy only agile enterprises having the capability to adapt continuously with the business environment will succeed.

To succeed in a competitive marketplace, an agile enterprise requires critical business intelligence to quickly anticipate, adapt, and react to changing business conditions. Unfortunately, many organizations striving to improve their agility soon discover that their legacy BI systems are not up to the challenge of the agile enterprise. As a result, their BI systems provide incomplete, unreliable, and outdated information to a limited group of analysts.

Where legacy BI systems fail to meet the broad and deep BI demands of the agile enterprise, a new evolution of critical BI systems provides superior flexibility, accuracy, availability, and timeliness. This new evolution of BI systems empowers BI stakeholders throughout the organization and enables them to act and respond more quickly to the immediate needs of the business.

Critical BI systems provide sustainable success in a dynamic environment by empowering information workers at all levels of the enterprise and enabling them to use actionable real-time information in their day-to-day tasks. Critical BI systems also empower application developers and IT professionals, enabling them to continuously and effectively create and manage BI solutions for the global marketplace. From a business perspective, critical BI not only translates into more cost-effective solutions, but also results in improved productivity and agility across the enterprise.

When agile enterprises design critical BI systems, they use a robust and dynamic BI platform capable of providing cost-effective BI solutions for the enterprise. Such BI platforms can deliver industry-leading BI technologies that uniquely satisfy the broad and deep demands of information workers, application developers, and IT professionals.

INFORMATION WORKERS

Such a critical BI system provides information workers at all levels of the organization with a single version of the data truth while also delivering data in a flexible accessible format.

Single Version of the Truth

From managers to analysts to front-line staff, a critical BI platform must be flexible enough to provide each information worker with the necessary information to anticipate and respond to the needs of the business. The BI platform must also be robust enough to ensure that the information is consistent and accurate throughout the enterprise. Consider the following examples of information workers who each use similar data in different ways.

- A marketing manager working to improve the organization's bottom line investigates the return on investment (ROI) of recent marketing campaigns to uncover opportunities for streamlining costs and improving overall marketing effectiveness.
- Heeding the manager's recommendations, a marketing analyst examines recent customer buying patterns to identify those specific customers who will likely take advantage of a newly launched product.
- With the new targeted marketing campaign in effect, a front-line sales representative recommends the new product to a targeted customer after perusing the customer's online profile and purchasing history.

Even though each information worker requires a different data view, the behind-the-scenes data must still be consistent and complementary to ensure that they are all looking at the same version of the truth. For example, the customer profile that the sales representative examines must be in sync with the list of customers that the marketing analyst has identified for the campaign. In addition, the buying patterns that the analyst views to produce the list of targeted customers have to be consistent with the data that the manager is examining to determine the overall ROI of the marketing campaign.

Without a single version of the truth, an organization will be mired in data conflict and will be forever arguing about which numbers are correct. This conflict not only creates general confusion, but it also slows down operational efficiency and may even result in legal ramifications if data quality standards violate compliance laws. From a business perspective, synchronizing the data across the enterprise may seem basic. However, many organizations really struggle with establishing a single version of the data truth and thus face large obstacles as a result of trying to bring together disjointed data systems and applying inconsistent business rules.

To help organizations remove these data obstacles, the critical BI system empowers information workers across the enterprise, from finance, sales, marketing, and executive management, with dynamic, consistent data views that support a single version of the data truth.

Accessible Information

Flexible data views are only valuable if they are presented in an easy-to-use format that is customized for each information worker. Consider the following examples of information workers with distinct data access requirements.

- A factory manager reviewing assembly line performance requires a dashboard that displays Key Performance Indicators (KPIs) and also uses easy-to-understand graphics, such as stoplights and gauges that quickly identify problem areas.
- An analyst spending a large part of the day crunching numbers to investigate the cause of declining assembly-line efficiency requires powerful report authoring tools to interact with the data quickly and efficiently without calling IT each time a new report is needed.
- Instead of using a specialized BI tool, a front-line support representative who wants to efficiently manage her time in responding to customer concerns about late shipments prefers to have customer data integrated into the applications that she uses on a daily basis.

From dashboards to self-service reporting to integrated reporting, the critical BI system empowers information workers and enables them to increase organizational agility by delivering relevant data in the format that provides them with the most value.

Application Developers

The critical BI system provides application developers with the ability to effectively integrate data from heterogeneous data sources, assemble that data into a meaningful analytic model, and then deliver the data in a variety of reporting scenarios.

Integration

When application developers build critical data integration solutions, they are challenged with assimilating large volumes of heterogeneous data in a short amount of time. To enable application developers to do their jobs effectively, the critical BI system provides the following capabilities:

Heterogeneous Data Access: Application developers can quickly access data from enterprise data systems, including mainframes, traditional databases, and external data providers.

Smart Data Cleansing: Data cleansing can be one of the most difficult and time-consuming tasks in data integration. Application developers require a variety of tools to cleanse and assemble the data for the business users. In addition, as part of data cleansing, application developers must ensure that any transformations are performed in an auditable manner to validate data quality.

Fast Data Processing: Across the enterprise, data volumes continue to increase as organizations accumulate large amounts of data about their customers, suppliers, employees, and products. To handle these volumes, application developers require a data integration technology that can scale in volume complexity.

The critical BI system provides application developers with a full-featured data integration engine and development environment to build high-performance data integration solutions. With access to a variety of data providers and smart data cleansing routines, the critical BI system enables the application developer to quickly solve simple and complex data integration problems for data sets of all sizes.

Analysis

When application developers create critical data analysis solutions, they are responsible for creating the back-end analysis architecture that provides information workers with a single version of the truth. To enable application developers to do their jobs effectively, the critical BI system provides the following capabilities.

Centralized Business Logic

To provide consistent data views to information workers, the critical BI system enables application developers to create a centralized data model that translates raw data into business terms, stores calculations, security rules, and any other business logic relevant to the organization. Within the model itself, data can be flexibly organized in a variety of ways to support the needs of sales, finance, marketing, and other areas of the enterprise. By using a common model, one information worker can analyze customer data by geography, and another information worker can analyze customer data by demographics with confidence that their results can be reproduced consistently.

Flexible Data Structures

Many application developers struggle with creating one analysis solution that efficiently supports both highly summarized data as well as detailed data views. With the critical BI system, application developers can easily combine aggregated and detailed data in the central data model. This integrated model will not only satisfy the information worker who wants to slice and dice sales data by geography, product, and time, but it will also provide the ability to transition from high-level analysis to a detailed analysis.

Advanced Analytics

To solve common business problems, application developers require a powerful modeling language that can efficiently handle the complexities of the business to produce meaningful financial, time series, variance, and statistical calculations. In addition, application developers may also need the ability to incorporate advanced data-mining logic into their analysis architecture to locate meaningful patterns in the data that users cannot find through traditional reports.

The critical BI system empowers application developers with a robust analysis engine to help them provide the central business layer to organize data. Within this central business layer, the application developer can create flexible user perspectives, rich calculations, and advanced data-mining solutions.

REPORTING

When application developers create critical data analysis solutions, they are responsible for creating the back-end analysis architecture that provides information workers with a single version of the truth. To enable application developers to do their jobs effectively, the critical BI system provides the following capabilities.

Rich Report Design

Application developers require a rich report authoring environment to create reports against multiple heterogeneous data sources, to organize data and calculations into tables, charts, and matrices, and to add report interactivity as necessary.

Flexible Information Delivery

Using the critical BI system, application developers can provide a wide range of reporting experiences to the enterprise. Information workers can slice and dice data in spreadsheets, interact with parameterized reports online or through e-mail, and even view reports in the native business applications they prefer. In addition, by fully integrating with the Web service platform, the critical BI system provides the capabilities to deliver information to any type of device, from desktops to laptops to handheld devices.

Self-Service Reporting

Self-service reporting provides information workers with the ability to create reports on their own without understanding the underlying database structures. Although self-service reporting provides a great amount of agility to information workers, application developers are responsible for assembling the data into a user-friendly reporting layer against which information workers can build reports.

The critical BI system empowers application developers with a flexible reporting architecture that provides self-service reporting as well as standard report delivery by using the format, device, and delivery schedule that is most appropriate for each business information worker.

IT PROFESSIONALS

A critical BI platform empowers IT professionals with the ability to quickly deploy integration, analysis, and reporting solutions, and to effectively manage these solutions over time with the utmost availability, reliability, and security.

Deployment

After application developers complete their data integration, analysis, and reporting solutions, IT professionals are responsible for preparing the solutions for information worker access. IT professionals must not only ensure that all necessary components are successfully migrated to the production environment, but they must also ensure that all required software is installed on information worker computers. The more efficient the deployment process, the quicker the turnaround time for new BI solutions and the easier to roll out periodic solution upgrades.

Through strong integration with Microsoft® Visual Studio®, the critical BI system empowers IT professionals with centralized deployment architecture to effectively release data integration, analysis, and reporting solutions by using automated and auditable procedures. In addition, to reduce the burden of installing reporting software on information worker computers across the enterprise, the critical BI system provides zero footprint reporting solutions that only require browser software on the client devices of choice.

Manageability

After deployment is complete, the IT professional is responsible for the ongoing management of BI solutions. To enable IT professionals to do their jobs effectively, the critical BI system provides the following capabilities.

Scalability: Provide integration, analysis, and reporting solutions that can scale in volume, complexity, and usage.

Availability and Reliability: Provide continuous BI to the global enterprise. With increasing demands for real-time BI solutions, the batch windows for processing data updates are becoming increasingly smaller at the same time that data volumes are increasing. Because critical BI is deeply integrated into the operations of the business, a BI system that is unavailable either because of system failure or because of standard offline data refreshing can seriously reduce the agility of the enterprise.

Security: IT Professionals are also required to administer the security of BI solutions. BI solutions often involve security at the data and report levels to ensure that each information worker only sees the subset of data for which she has been approved. Maintaining security is also a critical objective for IT professionals who must adhere to data compliance standards as well as uphold the data privacy of the enterprise.

CRITICAL BI FOR THE ENTERPRISE

The critical BI system sets the industry standard in providing critical BI solutions in the following ways.

- Uniquely empowers BI stakeholders: information workers, application developers, and IT professionals
- Provides a comprehensive, robust, and scalable BI platform to continuously create and manage data integration, analysis, and reporting solutions
- Maximizes ROI by providing the lowest total cost of ownership when compared with competitive platforms

With critical BI systems, information workers at all levels of the organization can effectively anticipate, adapt, and respond to the changing needs of the agile enterprise.

5

BUSINESS INTELLIGENCE: SOLUTION AREAS

Ignorance is the greatest threat to modern business. The risk of not knowing is immense. And, incomplete information can be even more harmful than no information, because we proceed and make decisions and act with conviction, falsely believing we know the true nature of the situation.

Business Intelligence (BI) describes taking data from its raw form and turning it into something usable on which business decisions can be based. It is an umbrella term that ties together other closely related data disciplines including data mining, statistical analysis, forecasting, and decision support.

Business intelligence allows businesses to leverage their information assets as a competitive advantage. It allows businesses to better understand the demand side of the business and manage customer relationships. It allows organizations to monitor results of change, both positive and negative. The move toward business intelligence is due to two major reasons: the information age and the economy.

In the information age, information is power. Enterprises that leverage, exploit, and maximize their information assets have a strategic advantage over their competitors. Business today moves at the speed of information. Getting the right information into the right hands at the right time is essential.

The other reason for the current emphasis on business intelligence is the economy. In an effort to survive the current economic storm, enterprises have focused on two main areas: reducing costs and increasing revenue. Where business intelligence systems have been implemented, an organization can pull together cost information from all internal organizations, which shows where the costs are, what the costs are, and provides a framework for making cost-cutting decisions. When the cost cutting is complete, the organization can view exactly what has been done and detail the impact in real-time.

In order to increase revenue, firms must focus on retaining customers as well as acquiring new ones. Retaining customers is always the most attractive. But understanding the profitability of the customer base is the first step. Typically, 20 percent of customers account for 80 percent of the profits. Firms first need to segment the customer base by profitability and act to retain the most profitable. Retaining these customers will provide the greatest lift to profits. Business intelligence enables this segmentation.

Business intelligence strives to eliminate guessing and ignorance in enterprises by leveraging the mountains of quantitative data that enterprises collect every day in a variety of corporate applications. The BI imperative insists that enterprises pledge to equip themselves with perspective.

Typically, business intelligence or OLAP (OnLine Analytical Processing) applications have most commonly been used by sales and marketing departments to analyze sales statistics. However, any data-rich business area needing to understand its operation more effectively can benefit.

BI APPLICATION AREA

BI area and classes of applications include:

- Market Analysis
  - Market Identification
  - Prospect Profiling
  - Market Segmentation
  - Promotion Planning
  - Marketing Performance Reporting
- Sales Reporting and Analysis
  - Sales Pipeline Reporting and Analysis
  - Channel Analysis
  - Sales Performance Reporting
- Finance Reporting and Analysis
  - Balance Sheet Reporting and Analysis
  - Accounts Receivable Reporting and Analysis
  - Accounts Payable Reporting and Analysis
  - Profit and Loss/Income Statement Analysis
  - Financial Budgeting and Forecasting
  - Cash Flow Analysis
  - Risk Management

- Supply Chain Analysis
  - Vendor Analysis
  - Purchase Planning
  - Cost Analysis
- Customer Relationship Analysis
  - Satisfaction Analysis
  - Churn Analysis
  - Customer Retention Planning

Other key areas include:

- Performance Management
- Compliance

BI APPLICATIONS

Examples of BI applications are:

- *Marketing and Sales Analysis:* The increase in data from sales and EPOS systems, as well as direct marketing campaigns, Web sites, and loyalty cards, means that the reporting and analysis requirements from this area often overwhelm the IT development resources available. A well-designed data mart can often provide a solution.
- *Budgeting:* Provides sophisticated custom roll-up facilities that can be used to consolidate and manage the budgeting process.
- *Financial Reporting:* Many organizations need to compare actual data with budget, last year, latest forecast, and so on. Using a multidimensional reporting application can often dramatically reduce the time required to create and distribute this type of analysis.
- *Customer Relationship Management and Quality Control:* Because incompatible systems have been implemented, organizations often find it difficult to analyze customer complaints and product defects, particularly if comparing them to totals sold or manufactured. Using data transformation services to load the different data sources into the data mart, this type of analysis suddenly becomes possible.
- *Web Site Click-Stream Analysis:* Huge amounts of data can be generated by Web site traffic. This type of application makes it possible to understand how visitors browse a site, which routes they take, and how often they return. Additional analysis may focus on peak traffic periods, or the effect of special Internet promotions on product sales.
- *Fraud Detection:* Widely used in health care, retail, credit card services, telecommunications (phone card fraud), and the like.

Fraud detection uses historical data to build models of fraudulent behavior and uses data mining to help identify similar instances.

- *Auto Insurance:* Detect a group of people who stage accidents to collect on insurance.
- *Money Laundering:* Detect suspicious money transactions (U.S. Treasury's Financial Crimes Enforcement Network).
- *Medical Insurance:* Detect professional patients and ring of doctors and ring of references.
- *Inappropriate Medical Treatment:* Detect patterns and fraud.
- *Detecting Telephone Fraud:* Telephone call model; destination of the call, duration, time of day or week. Analyze patterns that deviate from an expected norm.
- *Retail:* Analysts estimate that 38 percent of retail shrink is due to dishonest employees.

MARKET ANALYTICS

Market analysis provides visibility about an organization's markets, integrating information from sales, customer, and financial sources for a complete picture of the market. Business intelligence-enabled market analysis allows management to monitor markets, identify target markets, identify prospects, identify market segments, decide promotion strategy, evaluate the impact of promotions on prospects, analyze customer needs, analyze competitors, and plan marketing strategy.

Production or sales users can analyze prospect size to predict long- and medium-term revenue, purchase, and production figures. Business intelligence enables organizations to associate marketing data with financial, sales, and customer information to make informed strategic decisions to improve marketing effectiveness.

Key analysis areas are the following:

- Markets
- Prospects
- Segments
- Promotion effectiveness
- Competition

BI Applications

- Target marketing: Find clusters of "model" customers who share the same characteristics: interest, income level, spending habits, and the like.

- Determine customer purchasing patterns over time.
- Cross-market analysis.
- Customer profiling: Data mining can tell what types of customers buy what products (clustering or classification), assist in identifying customer requirements, identifying the best products for different customers, and use prediction to find what factors will attract new customers.
- Segmentation: Group customers into classes and class-based pricing procedures.
- Pricing: Set pricing strategy in a highly competitive market.
- Provide summary information: Various multidimensional summary reports, statistical summary information (data central tendency and variation).
- Competition: Monitor competitors and market directions (CI: competitive intelligence).

SALES REPORTING AND ANALYTICS

Sales forecast reporting and analysis provide visibility into an organization's sales pipeline, integrating information from sales, customer, and financial sources for a complete picture of sales performance.

Business intelligence-enabled sales forecasts allow sales management to monitor and act on individual opportunities, more accurately forecast current and future period revenues, and understand the drivers that distinguish won versus lost deals. Executives can use graphical dashboards to quickly access actual sales performance versus corporate targets and sales management forecasts.

Marketing users can analyze lead progression through each stage of the sales cycle to quantify the effectiveness and revenue impact of marketing efforts. Business intelligence enables organizations to associate sales pipeline data with financial, marketing, and customer information to make informed strategic decisions to improve sales effectiveness. Enterprisewide sales forecast reporting requires that each manager in the sales chain of command, each sales representative, each marketing and product analyst, and each executive have personalized access to sales pipeline information. Deploying reporting over the Web requires robust security to protect sensitive information and a scalable infrastructure to distribute reports to potentially thousands of geographically distributed sales force personnel.

Forecasting effectively through trend analysis and close probabilities assists sales managers and corporate executives who demand accurate revenue forecasts that blend up-to-the-minute sales cycle information with historically realistic close probabilities. BI provides easy-to-understand graphical and tabular forecast reports as well as statistical analysis to enable predictive modeling.

Executives and managers need to understand the underlying trends that create sales opportunities and result in deal wins and losses. These trends may vary according to geography, sales representative characteristics (tenure, experience, education, and quota achievement), competitor presence, and product. Business intelligence provides reporting flexibility to view sales pipeline data along the attributes of choice, while providing direct access to the deal specifics when needed.

Combining sales force data with product, customer, and marketing data helps sales and marketing analysts get access to information beyond what typically resides in SFA (Sales Force Automation) systems. Reports identifying product profitability, potential revenue by customer segment, and campaigns generating leads all require data from other corporate systems. BI integrates data from any number of corporate systems, allowing complete visibility into all variables affecting an organization's sales pipeline.

Key analysis areas are the following:

- Leads
- Pipeline
- Product sales
- Sales performance

BI Applications

Sales forecast reporting and analysis provides visibility into an organization's sales pipeline, integrating information from sales, customer, and financial sources for a complete picture of sales performance. Business intelligence-enabled sales forecasts allow sales management to monitor and act on individual opportunities, more accurately forecast current and future period revenues, and understand the drivers that distinguish won versus lost deals. Executives can use graphical dashboards to quickly assess actual sales performance versus corporate targets and sales management forecasts. Marketing users can analyze lead progression through each stage of the sales cycle to quantify the effectiveness and revenue impact of marketing efforts. Business intelligence enables organizations to associate sales pipeline data with financial, marketing, and customer information to make informed strategic decisions to improve sales effectiveness.

Secure reporting to multiple layers of sales management is important because enterprisewide sales forecast reporting requires that each manager in the sales chain of command, each sales representative, each marketing and product analyst, and each executive have personalized access to sales pipeline information. Deploying reporting over the Web requires robust

Table 5.1 Business Intelligence in Sales Reporting

BI Applications	Benefit
Leads Reporting and Analysis—Provides analysis of lead sources and trends	Better lead generation and management of lead qualification process
Pipeline Reporting and Analysis—Provides insight into opportunities and deals with the sales pipeline	Advanced monitoring of the entire sales pipeline and improved forecasting of sales
Product Sales Reporting and Analysis—Monitors trends in product sales	Increased sales and improved sales margins by targeting correct mix of products and services in the sales cycle
Sales Performance Reporting and Analysis—Delivers reporting and analysis of sales force performance	Constructive monitoring of the sales organization with ability to identify strengths and problems and take appropriate actions in a timely manner

security to protect sensitive information and a scalable infrastructure to distribute reports to potentially thousands of geographically distributed sales force personnel.

Sales managers and corporate executives demand accurate revenue forecasts that blend up-to-the-minute sales cycle information with historically realistic close probabilities. BI provides easy-to-understand graphical and tabular forecast reports as well as statistical analysis to enable predictive modeling (see Table 5.1).

Analyzing sales pipeline data from different business perspectives aids executives and managers who need to understand the underlying trends that create sales opportunities and result in deal wins and losses. These trends may vary according to geography, sales representative characteristics (tenure, experience, education, and quota achievement), competitor presence, and product. Business intelligence provides reporting flexibility to view sales pipeline data along the attributes of choice, while providing direct access to the deal specifics when needed.

Combining sales force data with product, customer, and marketing data aids sales, and marketing analysts require access to information beyond what typically resides in SFA (Sales Force Automation) systems. Reports identifying product profitability, potential revenue by customer segment, and campaigns generating leads all require data from other corporate systems. BI integrates data from any number of corporate systems, allowing complete visibility into all variables affecting an organization's sales pipeline.

SALES PIPELINE REPORTING AND ANALYSIS

Sales pipeline reporting focuses on the continual tracking and study of sales opportunities from raw lead to closed sale. Sales managers monitor key performance indicators to ensure marketing activities are generating the number, quality, and flow of leads necessary to meet sales targets and that sales cycles are progressing toward closure. Sales pipeline reports show the status and value of deals currently in the pipeline and probability-weighted forecasts. They also allow sales teams to drill to particular deal terms and conditions, and update material deal terms as negotiations occur. Business intelligence-enabled sales pipeline reporting not only generates and delivers these reports on a regular basis, but also allows managers to uncover competitive trends and performance anomalies by region, product, or customer segment.

CHANNEL ANALYSIS

Enterprises use a variety of channels to promote products to customers. Retailers use stores, catalogs, and Web sites to merchandise their offerings. Technology enterprises depend on direct sales forces, OEM partners, resellers, and systems integrators to deliver products to customers. Channel analysis allows organizations to understand the effectiveness of various sales channels, gauge channel growth, and compare channel margins. By reviewing channel comparisons and sales, channel and marketing managers can focus resources on building and maintaining the channels that best meet their customers' sales and service needs. Basic channel analysis includes sales and margin performance by channel, trends in channel utilization over time, and comparisons of channel performance by customer segment. More advanced analysis enables users to drill to individual channel partners, optimize pricing by channel, or match channel supply to channel demand.

COMPETITOR ANALYSIS

Sales forecast reporting is not complete without a review of competitive forces in sales cycles, their effect on revenue, and their implications for product development. Sales executives identify competitive win/loss performance at the region, sales representative, and individual sales cycle level. Basic competitive reporting measures competitive presence during the sales cycle and the success rates against each competitor. More advanced analysis enables users to drill into sales cycles of individual sales representatives to assess effectiveness in competitive situations, review transaction detail including products and options sold, or analyze discount programs used against various competitors.

FINANCIAL REPORTING AND ANALYSIS

Enterprises rely on business intelligence to provide visibility into the full scope of their business and financial operations. Finance and accounting departments across every industry are increasingly leveraging business intelligence to analyze sources of revenue and cost, view data underlying the statement of cash flows, and compare planned and actual revenue and expenses. The finance function has historically been charged with providing this fiscal transparency and has been increasingly leveraging business intelligence to analyze sources of revenue and cost, peer into the data underlying the statement of cash flows, and compare revenue and cost plans versus actual. Enterprises must satisfy public accounting requirements for government organizations and shareholders, as well as deliver financial data to financial analysts, business unit management, executives, and directors.

Externally, organizations must satisfy public accounting requirements for government organizations and shareholders. Internally, financial data must be delivered to financial analysts, business unit management, executives, and directors. In light of heightened government and shareholder scrutiny of corporate financials, finance and accounting departments are feeling additional pressure to make detailed information available more quickly and widely than ever before. Business intelligence uniquely provides efficient transparency, analysis, and delivery of corporate financial data.

BI is creating complete financial transparency for today's public and private enterprises alike. Security, widely distributed reporting, and out-of-the-box financial functions enable deployment success. As corporate financial data faces greater scrutiny, it is vital that finance and accounting departments make detailed information more quickly and widely available. Business intelligence uniquely provides efficient transparency, analysis, and delivery of corporate financial data.

Key analysis areas are the following:

- Profit and loss statements
- Balance sheet
- Cash flow analysis
- Revenue
- Costs/expenses
- Accounts payable
- Accounts receivable
- Planning and forecasting

BI Applications

Profit and Loss/Income Statement Analysis

The Profit and Loss (P&L) statement, detailing revenue and costs, is a useful tool for understanding the health of a business. Executives and managers continually monitor actual P&L versus budgets and forecasts to ensure their business is operating on plan. Automatically generated P&L reports allow managers to see current period revenue, costs, and profitability, period over period and business unit to business unit comparisons, and actual to plan. Business intelligence also permits development of more accurate and timely quarterly and yearly performance analysis, allows navigation to detailed revenue and cost components, and enables more refined margin and contribution assessment.

Cash Flow Analysis

A cash flow statement summarizes the operating activity of a business and provides insight into its ability to generate cash. Executives, treasurers, accounting professionals, and investors rely on cash flow statements to make financing, operating, and investing decisions. Standard financial systems, which are not optimized to handle large amounts of transaction-level data, take many hours to consolidate cash flow reports and cannot automatically distribute these reports. Business intelligence provides automatic, near real-time generation and distribution of the cash flow statement. Business intelligence seamlessly integrates sophisticated analysis including foreign currency exposure, optimized payables schedules, and cash flow predictions to ensure investments in short-term and long-term instruments are aligned with cash requirements.

Accounts Receivable and Accounts Payable Reporting and Analysis

Enterprises track Accounts Receivable (A/R) and Accounts Payable (A/P) to manage operating cash flow. Basic A/R and A/P reports include tracking the value of accounts that are 30, 60, or 90 days past due, monitoring distribution of receivables across customers, and reviewing payment trends for vendors across periods. More sophisticated analysis includes predicting potential bad debts, forecasting cash outlays, and tracking invoices and journal entries to the corresponding accounting representative. Businesses adding business intelligence to their A/P and A/R functions are benefiting from more efficient cash management.

Financial Budgeting and Forecasting

Financial budgeting and forecasting are an essential part of the business planning process. Executives and managers continually revisit forecasts as actuals are reported to determine how their business is performing relative to plan. Forecast reports allow detailed analysis by budget owners at every level of the organization and set the stage for determining sources of revenue and spending priorities. From simple reporting on actual performance versus budget to more sophisticated "what if" scenario creation and predictive modeling, organizations use business intelligence to make fact-based business plans and better monitor performance.

Risk Management

Credit risk, contract risk, currency risk, fraud risk, and audit risk are all becoming increasingly important components of risk that must be controlled by corporations and government agencies. Credit risk management, by combining data from publicly available sources with corporate customer and sales data, focuses on the determination of credit ratings, credit risk exposure in a particular country or industry, or the impact of increased customer defaults. Currency risk analysis involves evaluation of exchange rate exposure and predictive modeling of the impact of currency fluctuations on forecasted profit margin. Fraud risk analysis, through the use of statistical data-mining techniques, can uncover patterns of activity correlated with fraudulent events. Business intelligence, by enabling the use of data mining and statistical techniques for risk detection and management, empowers wider worker populations to help mitigate the various forms of corporate risk (see Table 5.2).

Supply Chain and Operations Reporting and Analysis

Supply chain analysis has emerged as one of the fastest growing business intelligence application areas. The proliferation of automated tracking systems, supply chain transaction systems, and electronic data interchange (EDI) systems have contributed to the rapid increase of data related to supply chain management. Enterprises in nearly every vertical industry need to have timely access to trends and indicators across key supply chain metrics.

A comprehensive supply chain analysis solution allows executives to analyze trends and details, quickly adjust inventory and distribution, identify vendor performance problems, and understand underlying supply chain costs and inefficiencies. These functions can work together to streamline supply chain system reporting, improve distribution and inventory methodologies, and improve communication of supply chain and vendor information.

Table 5.2 Business Intelligence in Financial Reporting

BI Applications	Benefit
Profit and Loss Statement Reporting and Analysis—Reporting and analysis of all revenues and operating expenses during a given period of time	Improved transparency into operating performance, ensuring that all personnel can access financial information relevant to their roles in the organization
Balance Sheet Reporting and Analysis—Delivers reporting and analysis of assets, liabilities, and equity	Unfettered analysis of assets, liabilities, and equity at any level of detail with transaction-level drilling for thorough investigative analysis
Cash Flow Analysis Reporting and Analysis—Tracks the inflows and outflows of cash	Improved analysis of cash inflows and outflows across operating, investing, and financing activities in order to optimize cash on hand
Revenue Reporting and Analysis—Delivers reporting and analysis of revenues at summary and detailed transaction-level information	Improved reporting on revenues in statutory reports as well as ad hoc analyses
Expenses and Costs Reporting and Analysis—Delivers reporting and analysis of costs and expenses across multiple business dimensions for any given time period	Reduced cost outlays by activities, employees, and business units across the organization
Accounts Payable and Accounts Receivable Reporting and Analysis—Enables analysis of current and aging accounts, broken down by days outstanding, business organizational structure, and vendor or customer	Improved tracking of Accounts Receivable to increase cash on hand Improved tracking of Accounts Payable for monitoring all liabilities

Key analysis areas are the following:

- Inventory
- Vendor performance
- Distribution network efficiency

Web Site Reporting and Analysis

Enterprises study Web site visitor activity to understand customer and prospective buyer interests, improve Web site design and navigation, and increase the profitability of the online sales process. By tracking and analyzing Web traffic patterns, site navigation, page views, and transactions, online marketers can design more compelling Web sites and give visitors better incentive to buy and return. With the low switching costs present on the World Wide Web, enterprises know that a visitor's experience must offer enough value to induce repeat visits. Enterprises rigorous in their analytical approach to online merchandising and sales are realizing lower marketing costs, higher online revenue, and improved customer loyalty.

Key analysis areas are the following:

- Inventory
- Vendor performance
- Distribution network efficiency

Product Sales Reporting and Analysis

Product sales analysis, most commonly used by retailers, enables the continuous monitoring of point-of-sale data to uncover sales trends, investigate product demand, and optimize merchandising strategies. Various levels of analysis, from summary reporting to statistical trending, are required by executives, store managers, product managers, and marketing analysts, as well as external suppliers who provide materials or finished goods. Business intelligence makes sense of the growing volume of transactional data by identifying trends and opportunities that create competitive advantage for enterprises that know and understand their sales drivers.

Key analysis areas are the following:

- Inventory
- Vendor performance
- Distribution network efficiency

Corporate Performance Management (CPM)

Enterprise business intelligence is also the key enabler for Corporate Performance Management (CPM). CPM allows enterprises to proactively monitor and eliminate impediments to performance before they affect financial results. CPM is a real-time initiative that aligns everyone in the organization in support of a business strategy by proactively monitoring and measuring business performance and alerting key personnel at all levels when problems occur in time to take corrective action.

Successful CPM initiatives require a flexible and reliable reporting environment that unifies everyone from the CEO to a mailroom clerk to quickly resolve performance problems. CPM applications create shared accountability throughout an organization by different levels of reports pertaining to a particular performance area with varying levels of detail. For example, a senior executive may receive an update on days of inventory as part of a dashboard of high-level key performance indicators she views daily; simultaneously, the inventory manager receives a report showing the exact status of each piece of inventory with the ability to drill down to determine what might have caused a sudden change in inventory (e.g., a large number of returns or cancelled orders). The manager can investigate and take corrective action to rectify the problem. However, what if the senior executive is particularly concerned and wants to check throughout the day to see if there is a change. She has the option of getting a just-in-time update by performing the query in real-time directly from the dashboard.

CPM exemplifies the power of BI in transforming an enterprise into a sense-and-respond environment based on improved information flow. Using BI to effectively measure overall business performance—as well as performance of specific functional areas or lines of business—will play an important role in creating a culture of measurement and accountability (see Figure 5.1).

Methods of managing performance include the following:

■ Visualize key performance data effectively in graph format.
■ Convey performance results quickly with visuals.
■ Use scorecards and dashboards as gateways to first-order and advanced analysis.
■ Monitor red zones and define threshold levels to set indicators and trigger alert deliveries.
■ Link individual KPIs (Key Performance Indicators) to corporate goals.
■ Cascade scorecards and dashboards throughout the organization and across the value chain.
■ Deliver scorecards and dashboards via e-mail, on a scheduled or alert basis.
■ Reach all individuals, from executive managers to new associates.
■ Incorporate all enterprise data—financial and operational—from every business process worldwide.
■ Access personalized secure scorecards and dashboards from any Web browser, intranet/extranet, or portal.

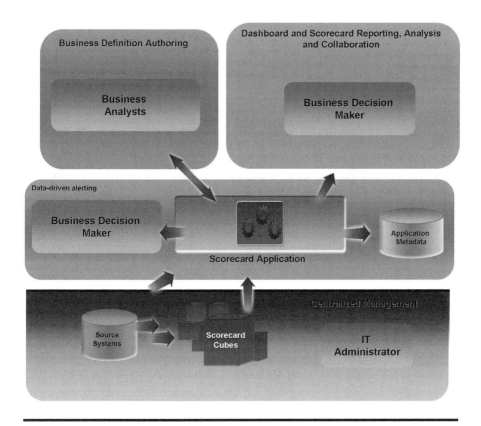

Figure 5.1   Scorecard architecture

COMPLIANCE

The demand for timely reliable information is placing a tremendous amount of attention on corporate reporting and auditing practices. Financial institutions must comply with an ever-increasing number of regulatory mandates that affect both customer and capital information, such as the U.S.A. Patriot Act requirement to gather as much customer information as possible at all levels of interaction and the Basel II directive to track capital adequacy across the enterprise.

In addition, new legislation such as the Sarbanes–Oxley Act has brought financial reporting and forecasting to the forefront of today's executive agendas. At the same time, industry-specific regulations make unique demands for reporting and compliance in just about every public and private sector, from education to banking to pharmaceuticals.

Some types of compliance affect all enterprises in all industries, such as the Sarbanes–Oxley Act. Created in response to high-profile accounting and management scandals, the Sarbanes–Oxley Act is crucial to restoring investor trust in public markets and reestablishing corporate and accounting credibility. The reforms are designed to reduce fraud and oversights in corporate reporting, accounting, and auditing practices.

The gist of Sarbanes–Oxley is that there are many more hoops that corporations must jump through when preparing and presenting their financial statements. Among other things, companies must set up new, more stringent business governances, as well as policies and processes for collecting, certifying, and disclosing financial information.

Financial reporting and forecasting are placed in the spotlight by the Sarbanes–Oxley Act. Facing tightened quarterly and annual reporting deadlines, corporations must gather disparate information from throughout the enterprise, then quickly process, format, and distribute it. Corporate executives are held personally responsible for the accuracy and consistency of information located anywhere in their organizations.

Given just 48 hours to report material changes in financial status that could affect the stock market, these senior officers need to be able to quickly summarize financial information and drill down to the underlying details. They must establish and certify internal financial controls and be able to open up their audit trails to government auditors. Criminal penalties have been established for altering documents, which means enterprises must implement auditable document retention procedures and schedules.

The filing deadlines have been shortened, forcing corporations to produce complex financial statements faster than ever before. Disclosure of additional off-balance-sheet transactions (such as debt leases, lines of credit, and guarantees) must also be included in an organization's regulatory filings, providing a more complete picture of financial health and making potential financial problems more visible.

The Sarbanes–Oxley Act may significantly affect telecommunication companies. Telecom firms may also be subject to a higher standard of conduct after the relatively large number of recent industry-related scandals. Sarbanes–Oxley, which contains requirements for real-time information, has put enhanced pressure on organizations and their focus now is on achieving the following financial reporting criteria:

- Aggregating financial data
- Making financial details more accessible
- Enabling frequent flash reporting
- Drilling down on accounting reports
- Highlighting key analysis areas based on tolerances and financial metrics
- Segmenting reporting into material/significant elements
- Enabling workflow
- Instilling a financial management mindset within the organization

Corporate Reporting and Internal Governance

Apart from any formal legislation or government mandate, every organization has myriad reporting issues with which to contend. Managers cannot wait until the end of the month to evaluate strategy decisions or compute corporate performance gains. They want to know where the business stands at any given moment so they can proactively satisfy customers and meet corporate goals.

The job of satisfying these reporting requirements often falls on the shoulders of Information Technology (IT) professionals, who typically rely on business intelligence software to lend cohesion to a diverse information landscape. When properly deployed, enterprise business intelligence and reporting tools can deliver information in a variety of ways to many types of people, both inside and outside the enterprise.

Financial professionals, in particular, from CFOs to controllers to analysts, need to be able to access and analyze real-time information from multiple business units to obtain a complete picture of corporate performance. CFOs need to develop systems that can collect, organize, and disseminate information, with the goal of providing true visibility into the operational drivers of share price and the unexpected events that affect valuation.

This implies unique needs, for example, the ability to dynamically read varying hierarchies such as charts of accounts. Dynamic posting allows financial users to generate new summary statements from individual detailed reports, as well as to transfer balance sheet items across years. Embedded rollup routines use stored, dynamic, or combined reports to simplify rounding issues, and fully integrated OLAP capabilities streamline the transition from summary reporting to analytical processing.

Real-time data sharing between financial and related nonfinancial processes, such as transaction-processing and Enterprise Resource Planning (ERP) systems, ensures data consistency and accuracy, allowing managers to consolidate interorganization balances and business unit totals prior to reporting, shortening the close cycle.

In addition to reporting, most enterprises need fast, cost-effective data integration to create accurate snapshots of financial performance. An auditable, single-step process can load consolidated data, eliminating the possibility of error that multistep processes introduce.

Other Regulations

There are hundreds of new regulations adopted globally to create safeguards against financial, virtual, and physical crimes. Some of them are discussed below.

The U.S.A. Patriot Act

The U.S.A. Patriot Act, passed in October of 2001, is a vital tool in the continuing effort to prevent future acts of terrorism. The primary goals of the act, which is also known as the Uniting and Strengthening America by Providing Appropriate Tools Required to Intercept and Obstruct Terrorism Act, are to gather and cultivate detailed terrorism-related intelligence so that potential terrorist plots can be detected, disrupted, and prevented, and to build our country's long-term counterterrorism capacity. Although most of the act is targeted toward law enforcement agencies, it also poses a significant impact on financial organizations, higher-education institutions, transportation and logistics enterprises, as well as any foreign organization doing business within the United States.

Integrated Justice and Law Enforcement

The act significantly increased the surveillance and investigative powers of law enforcement agencies in the United States. Above all, the act stresses the importance of having immediate access to timely, accurate, and complete information that can be quickly exchanged among law enforcement and justice officials. Rapid reporting and information sharing is essential not only to respond to threats of international terrorism, but also for domestic terrorism, natural disasters, and many routine incidences related to public safety.

Money Service Businesses—The Bank Secrecy Act

Title III of the U.S.A. Patriot Act requires all registered financial brokers and dealers (also called Money Service Businesses, MSBs) to implement anti-money-laundering programs designed to achieve compliance with the Bank Secrecy Act (BSA). Money laundering involves acts committed to conceal or disguise the criminal origin of funds so that the unlawful proceeds appear to have derived from legitimate sources. All financial institutions, including credit unions, are subject to recordkeeping and reporting requirements of the Bank Secrecy Act and the Treasury Department's regulations. These organizations need BI tools to file Suspicious Activity Reports that help law enforcement officials identify fraudulent, criminal, or terrorist activity.

Financial Services—Basel II

Financial institutions are in the business of taking risks for their customers and are therefore subject to a high level of industry supervision and control. As part of Basel II, a directive designed to harmonize international banking regulations, government auditors demand proof of robust compliance controls and enhanced levels of management oversight, forcing senior

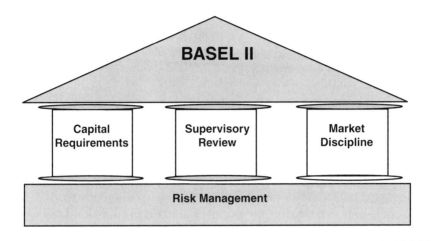

Figure 5.2   Basel II compliance

managers to coordinate and monitor regulatory compliance activities. This intensive regulatory environment demands an enterprise view of risk that enforces common reporting structures across the enterprise and the development of information sharing across functional departments.

Key technologies to support an enterprise risk-management framework include reporting, data collation, data cleansing, exception monitoring, and compliance reporting. A complete BI toolset provides the requisite flexibility for dealing with reporting period changes and discrepancies that result from mergers and acquisitions, including direct drill-through to daily postings from aggregated data in a data warehouse.

Being able to access live data throughout the enterprise makes it much easier to obtain correct data. Most BI environments include a graphical user interface to simplify the development of new reporting applications and a dynamic chart of accounts to produce "self-maintaining reports," with no need to reload the data warehouse just because changes are made.

Gramm–Leach–Bliley Act

The Gramm–Leach–Bliley Act (GLB Act), also known as the Financial Modernization Act of 1999, is a federal law enacted in the United States to control the ways that financial institutions deal with the private information of individuals.

Can-Spam Act of 2003

The Can-Spam Act of 2003 is a commonly used name for the United States federal law more formally known as S. 877 or the Controlling the Assault of Non-Solicited Pornography and Marketing Act of 2003. The law took effect on January 1, 2004. The Can-Spam Act allows courts to set damages of up to $2 million when spammers break the law. Federal district courts are allowed to send spammers to jail and triple the damages if the violation is found to be willful.

The "Do Not Call" List

The "Do Not Call" list is a registry of phone numbers in the United States that telemarketers are prohibited from calling in most circumstances. The list is maintained by the National Do Not Call Registry of the Federal Trade Commission (FTC), and consumers can contact the agency to have their numbers registered. Organizations are prohibited from making calls to sell goods or services to any numbers listed, and are subject to substantial fines if they fail to comply.

HIPAA

HIPAA is the United States Health Insurance Portability and Accountability Act of 1996. HIPAA seeks to establish standardized mechanisms for Electronic Data Interchange (EDI), security, and confidentiality of all healthcare-related data. There are two sections to the act. HIPAA Title I deals with protecting health insurance coverage for people who lose or change jobs. HIPAA Title II includes an administrative simplification section that deals with the standardization of healthcare-related information systems.

Role of Business Intelligence in Achieving Compliance

A good BI system will be needed to ensure the timely and accurate analysis of data needed to be in compliance with Sarbanes–Oxley and various other regulatory requirements. Of course, finance is just one piece of the corporate governance puzzle. There are many good reasons why enterprises wish to standardize the way they produce and share information. Having a unified reporting layer ensures consistency in the way information is accessed, delivered, presented, and stored. Additionally, it can dramatically reduce the expenses associated with deploying and maintaining the environment.

Most organizations deliver financial closing statements using a detailed consolidation cycle. Data from throughout the enterprise must be manually manipulated to create flash sheets and financial statements. Financial reporting is very complex. Rather than the usual columnar reports, it is cell oriented. Many analysts use spreadsheet programs such as Microsoft Excel to compute results and calculate totals because they help automate this process.

However, to meet today's more stringent reporting requirements, these disparate financial reporting activities need to be combined into automated repeatable processes. This involves simplifying data integration, report generation, and information delivery.

Historically, financial reporting software was mainly geared to analysts and power users trained to use multidimensional databases, OLAP tools, and other specialized software. The latest business intelligence tools need to simplify the process considerably with user-friendly financial reporting capabilities that automate the collection, calculation, analysis, and presentation of financial information. Forecasting capabilities enable analysts and other financial professionals to quickly calculate moving averages, linear regression, and exponential moving averages on measures that they select from drop-down lists. There is no need to retrieve, group, and calculate data by columns, or to sort it numerically or alphabetically. Reports need to be easily sorted in order of liquidity or according to mandated rules. Excel PivotTables can be automatically generated and saved from any report, combining advanced reporting with Excel's powerful data manipulation capabilities.

Different users have different needs. Financial analysts require robust analysis tools. Production managers want to receive alert-based reports, meaning they want to be notified when certain conditions change. Sales managers want to receive periodic summaries, yet have the ability to drill down into the data when something piques their interest. Line-of-business managers need action statements that allow them to set other activities in motion. Executives want to be notified via pagers or cell phones when certain thresholds are reached.

An effective BI environment must be able to cost-effectively accommodate all of these scenarios. The best BI tools can maintain drill-downs and styling when a report is saved to an Excel or PDF file, and perform columnar or interrow calculations. These advanced financial reporting capabilities allow analysts and other business users to spend more time planning and forecasting, and less time validating, justifying, and auditing. As a result, enterprises can close each quarter faster, more accurately, and with greater understanding, and corrective action can be taken quickly, if necessary.

Some of the features organizations need are the following:

■ An easy-to-use portal interface, enabling executives and other financial professionals to track key performance indicators to measure the health of the business at any point in time
■ The ability to bring data directly into Excel or save reports as Excel files, providing financial professionals with a familiar tool while preventing data-entry errors

- Easy but powerful data analysis and the ability to drill down to real-time posted transactions (ideally, while in Excel) giving analysts the ability to quickly and simply research problem areas in detail
- Ad hoc reports that let managers discover and immediately communicate changes in the business
- A built-in audit trail that gives independent auditors an in-depth view of financial activity at any point in time
- Auditable report management that enables financial professionals to prove compliance if questioned
- Role-based security to ensure the proper flow of information throughout and beyond the organization
- Real-time reporting (within 48 hours) of material events that could affect an organization's financial performance, with the ability to consolidate data at report time
- Access to all data in the enterprise
- A common view of information to facilitate collaboration among analysts, financial professionals, and high-level corporate executives
- The ability to deliver information to any appropriate party, within or outside the organization, in any electronic or print format

TOWARD ENTERPRISE BI

Both public and private sector organizations have become highly proficient at capturing large volumes of data, with the capability to hold information about every customer that has ever transacted business with them. Unfortunately, for many such organizations the good news ends there: although they have been working diligently at taking in information and building up terabytes of stored data across a variety of operational systems and databases, the one vital thing that they have failed to improve upon is the ability to make more effective use of that data.

The era of the isolated business intelligence tool is drawing to a close. Certainly, most of the latest product releases from the industry's leading players have been targeted toward the delivery of enterprise BI services, and much of the recent consolidation activity can be attributed to a movement toward an "all-in-one" strategy for BI. All organizations need to become more efficient in the way that they exploit their data assets.

In the BI arena, data has no intrinsic value unless it can be used to support business decisions. Business will only be able to improve its information services, and obtain real value from the ever-increasing data silos that it continues to generate, when it accepts that there are significant advantages to be gained from integrating and standardizing its approach to the management of BI services (see Figure 5.3).

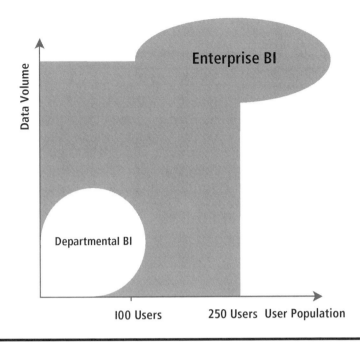

Figure 5.3   Enterprise BI and departmental BI reach

Visible cost savings that come from BI product consolidation will accrue from the simplification of systems management and systems support infrastructures. At the same time, Butler Group believes that the real financial benefits to organizations will come from the better, more consistent, and more competitive use that the business can make of integrated operational intelligence.

BI has reached a crossroads, and its value to business will only be improved when products are delivered that are capable of being used as genuine enterprisewide, intelligence-led, data access, management, and information delivery solutions. Any organization that cannot deal with data quality and consistency issues from within the confines of its BI platform does not have a credible product in place, nor does it have a plausible BI strategy. Ultimately there needs to be a high priority set on BI to take first-line responsibility for all the data that it services. Accepting poor and unproven data into enterprise decision-making systems is in Butler Group's opinion the ultimate recipe for disaster.

Today businesses are strongly inclined to find a better, more efficient, and more cost-effective approach to utilizing the competitive and commercial value of their data assets. The BI industry itself is eager to provide its services at a cross-organization level through the deployment of strategic and enterprise-pervasive BI, but the business community seems somewhat reluctant to acquiesce.

Therefore, there is a need to clearly spell out the incremental value proposition that the extended enterprise use of BI can provide. The future value of BI to business organizations will come from the extended use of enterprise intelligence services that incorporate the use of products with the capacity and capability to be used as genuine, enterprisewide, intelligence-led, data access, management, and information delivery solutions. Business decision makers should be challenging their technology suppliers to deliver intelligence-led solutions that can deliver pervasive BI facilities that empower information users at all levels within the organization to work smarter and in a manner that makes the organization more efficient, more agile, and ultimately more operationally competitive.

6

REAL-TIME BUSINESS INTELLIGENCE

Timely, relevant information has become the scarce commodity. For the workforce, intelligently managing and distributing information offers real opportunities for cost and time saving, improved customer servicing, and higher revenues.

It has become evident in recent years that efficient and timely communication is a key to progress in all fields of human endeavor including business. However, present communication methods are totally inadequate for future requirements. Information is now being generated and utilized at an ever-increasing rate because of the accelerated pace and scope of business activities, at the same time the growth of organizations, increased specialization, and segmentation have created new barriers to the flow of information.

There is also a growing need for more prompt decisions at levels of responsibility far below those customary in the past. Undoubtedly the most formidable communications problem is dealing with the sheer bulk of information. In view of the present growth trends, automation appears to offer the most efficient methods for retrieval and dissemination of this information. During the past decade significant progress has been made in applying machines to the processes of information retrieval.

Automatic dissemination has so far been given little consideration. However, unless substantial portions of human effort in this area can be replaced by automatic operations, no significant over-all improvement will be achieved. Even the information retrieval processes mechanized so far still require appreciable human effort to organize the information before it is entered into machines (see Table 6.1).

TABLE 6.1   Traditional View of Business Intelligence

Make Better Business Decisions
Increase organizational credibility
Gain timely and accurate insight into business operations and processes
Regulatory Compliance
Patriot Act, Bank Secrecy Act / Anti-Money Laundering Act
Sarbanes–Oxley, Basel II
Optimize Operational Efficiencies
Information Analysis for LOB and business units
Real-Time Data Integration
Data as asset
Integrated Data = Efficient + Intelligent Business

FIRST GENERATION BI—LEARNING FROM THE PAST

The first generations of Business Intelligence (BI) solutions were departmental and attempted to eliminate very specific information gaps relating to a functional area or line of business, which has created a fragmented BI landscape across most enterprises. Without a unified BI environment, it is difficult for enterprises to roll out large-scale applications that take critical information buried in operational systems and deliver it in real-time to anyone who needs it, regardless of where they are. This situation has also made it difficult to keep pace with rapidly changing business requirements.

First-generation BI solutions have also relied exclusively on enterprise Data Warehouse (DW) implementations. DW implementations, like ERP (Enterprise Resource Planning) implementations, have been costly and have suffered delayed solution paybacks as well. The data warehouse plays an important role in enabling BI and reporting applications, by protecting operational systems when necessary and creating a comprehensive and accurate view of information locked in packaged applications. However, BI and reporting solutions that are based solely on the data warehouse contain too much information latency to satisfy rapidly emerging requirements for real-time information.

Organizations have taken similar approaches to filling in the business-process white spaces across the enterprise. Like BI, first-generation enterprise application integration efforts were quick fixes that addressed a business-processes inefficiency that pertained to a couple of different systems. This approach was limited for several reasons.

Organizations "hard-wired" these systems together with custom-coded integrations that were generally expensive, fraught with risk, and impossible to maintain once completed. There was an incremental improvement but once the initial solution was complete, there was not much opportunity to innovate further because the code was rarely reusable for subsequent projects. As a result, most enterprise architectures evolved from a series of siloed packaged applications to an even more complicated tangle of "spaghetti-code solutions" that initially decreased the size of the white spaces, but failed to eliminate them. These solutions simply lacked the flexibility to orchestrate a process that spanned multiple systems and people to complete a business transaction and could not evolve to support continually optimized processes based on changing conditions.

The failure to deploy global solutions was largely due to technical constraints. Enterprises turned to these incomplete solutions because they were the best option available until Internet connectivity and interoperability standards emerged in the mid-1990s. However, Internet technologies, which enabled the wide dissemination of information and transparent systems interoperability, did not mature to support large rollouts until the end of the 1990s. In most cases, technology vendors had not evolved their products enough for enterprises to take advantage of these standards or, by the time mature, Internet-based products were widely available, the economy had turned south, and enterprises had cancelled most projects.

Yet it remains imperative for enterprises to seize the opportunity that now exists to transform their packaged applications for competitive advantage without succumbing to previous pitfalls.

A NEW MINDSET—INFORMATION NOW

Shortening the time to finding the right information is the guiding principle of the Information Now Mindset. As more businesses strive for the ideal of the real-time enterprise, there is growing interest in reducing the latency of BI delivery. Making faster decisions based on more real-time information can benefit enterprises seeking faster and more efficient operational processes.

In the past, people were willing to wait days, weeks, or even months for important business information. Waiting for a clerk to sift through reams of documents only to uncover the wrong information and to repeat the process again and again was considered a practical limitation as there was not any choice, and it didn't really matter because everyone had to wait the same amount of time for the information they needed. Today, however, most of us cannot afford to wait around, and we have all kinds of technology at our disposal to make sure that we do not have to.

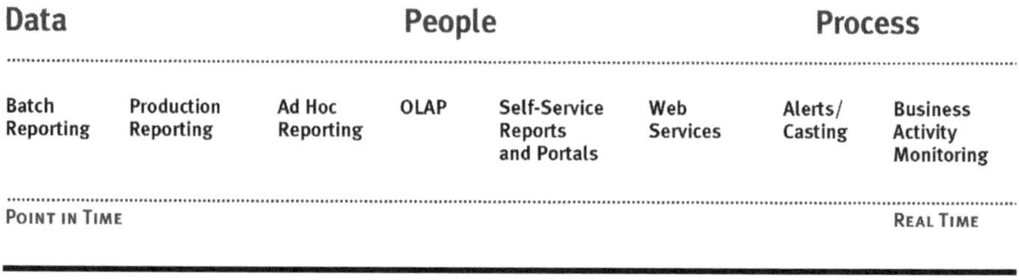

Data				People			Process
Batch Reporting	Production Reporting	Ad Hoc Reporting	OLAP	Self-Service Reports and Portals	Web Services	Alerts/ Casting	Business Activity Monitoring
POINT IN TIME							REAL TIME

Figure 6.1   From Point in Time to Real-Time

The Value of Time

Success in the present business environment is often dependent on the ability to access information, analyze it, and make a decision in time to outmaneuver a competitor (see Figure 6.1). Modern communication technologies with their common standards and universal connectivity, the Internet, has led to an explosion of information across the enterprise and the extended enterprise and created an infrastructure with a low-cost omnipresent user interface to power the "information now" world.

Unlike in the past century, enterprises cannot wait to get the information needed to run a business. Enterprises want to do business in real-time. The real-time enterprise has emerged as a new computing paradigm that lets organizations create flexible systems architecture to accommodate a dynamic business environment.

The push to build the real-time enterprise is based on one central premise that is true in any economic climate: information is valuable, and the value of information increases as more people across an enterprise can gain immediate access to it and use it to make informed decisions that improve business results.

Real-time information allows businesses to immediately uncover and drive out hidden costs, identify and correct business performance problems before they become catastrophic, and bring about changes in behavior in individuals and groups that improve accountability and better align them with corporate goals.

Expanding the Business Intelligence Framework

Making timely business decisions has never been easy, but today it is more difficult than ever, in large part because of the increasing volume of available information. Business managers face several challenges:

- They want to base their decisions on the full range of information available, but that range is growing exponentially as computer systems become more powerful.

- They rely on ever-more-sophisticated analytical tools, which increase the challenge of maintaining real-time performance.
- They are moving toward service-oriented architecture with flexible information infrastructure, a major benefit that brings with it a serious side effect: user queries and report requests are harder to predict.

All these challenges drive organizations to improve the performance of their business intelligence systems. Modern business intelligence solutions address the real-time information requirements for all of the key business technology initiatives cited by executives from different industries and geographies, time and again, that is, business agility, cost reduction, and visibility into key performance metrics to name a top few.

Traditional BI

At the core of most BI systems are software products providing query, reporting, and analysis functionalities, sometimes referred to as OLAP (OnLine Analytical Processing). Traditionally, these functions have been based either directly on operational systems, or more commonly on a data warehouse or data mart, which assembles and restructures data from one or more operational data stores. The result of the data combination can be stored in another data store or can be virtual. In many cases, the data are obtained from both relational and multidimensional OLAP schemas, and are combined into a Hybrid OLAP (HOLAP) which is usually virtual. Web services can be used to make this functionality available remotely over the Internet.

Closed-Loop BI

A number of BI vendors offer Web service front ends to their basic OLAP products; however, the remote services merely provide access to a set of predefined reports, or generate predefined management alerts, with no ability to perform analysis dynamically. This limitation applies to many so-called portals or dashboards, in which the service can be understood simply as a one-way information flow from the data store to the manager. Even with full functionality for query, reporting, and analysis, the traditional view of business intelligence only gives a partial picture of the process, lacking a broader system purpose and context of management control and action.

The traditional approach encourages a misleading view of business intelligence as a passive activity by individual managers. It is a known pitfall for business intelligence to be taken over by clever number crunchers, identifying fascinating statistical patterns with no practical relevance for management action. However, business intelligence can be used as a tool to process and interpret information which then can be used to act upon and monitor the effects of their actions. If the actions have the expected effect on business performance, this helps to confirm the original interpretation, otherwise if management intervention does not work in the expected way, then this should trigger further analysis. This management feedback and learning loop is a key element of true business intelligence where BI works as a closed control loop.

Closed loop business intelligence also includes the possibility of management actions whose primary purpose is to gain more information or intelligence: to learn something or to test a hypothesis.

Real-Time Business Intelligence

In the real-time enterprise, transactions and events can be intercepted and analyzed as they happen, without waiting for them to reach a data store, and this leads to the possibility of a real-time or near-real-time system response.

Real-time business intelligence (see Figure 6.2) removes information blind spots to improve real-time decision making by everyone in a business, regardless of function, level, or location, by taking data buried

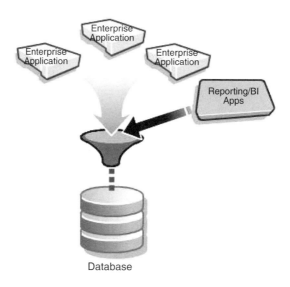

Figure 6.2   Real-time business intelligence

in operational systems and delivering it to people exactly when and how they need it. This inevitably improves responsiveness, accelerates business cycles, and increases profitability.

Real-time solutions help enterprises realize better performance and accountability, comply with legal mandates and industry standards, effectively collaborate with customers and partners, and create a platform for strategic growth through mergers and acquisitions.

OPERATIONAL BI—A BUSINESS IMPERATIVE

Organizations are facing intensifying global competitive pressures, customer demands for better service, investor demand for increased profitability, market demand for better governance, and the need to comply with more demanding government regulations. These pressures are forcing them to ensure that they know what is going on in their business on a daily basis to react quickly to rapidly changing conditions.

Business intelligence systems are used by business users to increase revenues, reduce costs, and otherwise optimize business operations. Users typically work with a set of tools such as dashboards, scorecards, reports, and analyses to derive business insight and improve the decision-making process in terms of speed and quality. Data Integration (DI) assumes a very important role in a BI system, by making the required information available for the end-user tools. Common data integration products include data warehouses, operational data stores (ODS), or replicated databases to support offline reporting.

Traditionally, BI systems were used for supporting decision-making processes that are strategic in nature, where the decision-making timeframe can span months (e.g., planning) or weeks (e.g., program or campaign management). Today, enterprises are faced with competitive pressures and customer demands that require faster responses. These needs call for using BI to improve decisions and business operations on a daily basis and support "same day" decision cycles (e.g., customer service, supply chain management, logistics optimization). This tactical usage of BI systems is called operational BI.

Operational BI goes beyond traditional BI implementation in two key areas:

- Expanding the use of BI to virtually anyone (information democracy)
- Access to the most current information for decision making.

This demands that the technology and architecture that support BI systems be pervasive, productive, and efficient. In addition, these BI systems need to ensure on-demand or right-time information availability.

There are three fundamental building blocks for an operational BI system:

1. *Information Delivery:* Efficient, pervasive, and productive delivery of information to the user
2. *Information Serving:* A database platform that makes information available for delivery
3. *Information Integration:* Real-time, on-demand (or just-in-time) integration

Operational BI means delivering more information to more people in support of tactical as well as strategic decision making. This requires solutions to be productive, simple, and ubiquitous. Tools such as dashboards, reports, and portals need to be seamless and intuitive for all users. Furthermore, solutions must lend themselves to rapid changes, customization, and self-configuration to allow for adaptation to market needs as well as personal needs and preferences.

TWO TO TANGO—BUSINESS PERFORMANCE MANAGEMENT AND REAL-TIME ENTERPRISE

Winning in ever-more volatile, complex markets, managers today must make faster and smarter decisions than their competitors. Accurate, timely, and actionable information along with sophisticated decision support tools are required. Winners in today's multifaceted markets will utilize business intelligence for better performance management and to act in real-time to achieve competitive advantage. Competitive advantage will go to those who can harness data for better business intelligence and decision making.

No matter how smart the management team, if they are working with dated or bad data, decisions will be either wrong or delayed. In many cases, due to siloed systems, justified as providing individual groups with the speed and flexibility to make better decisions, decision makers are using different versions of what should be consistent data. External factors such as price volatility and stringent regulations, coupled with an explosion of data available to managers, has made the use of patchwork applications and spreadsheets an unsustainable system environment for organizations looking to compete in a fast-paced marketplace.

Today business intelligence is considered one of the most strategic software investments an organization can make. The right decision support tools can harness data for better business results. The two most compelling trends observed across the enterprise technology landscape are Business Performance Management (BPM) and real-time data warehousing.

BPM is gaining momentum as businesses become more focused on the bottom line and the cost and benefit of every function. At the same time, the pace of business is speeding up as customer demands become more intense and competitors move more quickly than ever to meet their needs. Combining these two trends leads to the need for real-time business intelligence solutions and hence Real-Time Enterprise (RTE).

Business performance management and real-time enterprise techniques now being developed will greatly contribute toward achieving enterprise goals. They will enable organizations to face the challenges of the new economy by making it possible to accept information in its original form, disseminate the data promptly to the proper places, and furnish information on demand in a format easily understandable as well as ready for decision making.

Business Performance Management (BPM)

Business performance management is a set of processes and methods that help organizations optimize business performance. "Business performance management" is a term generally used to describe a set of concepts to improve business decision making by using fact-based support systems.

BPM, seen as the next generation of business intelligence, is focused on business planning and forecasting and involves consolidation of data from various sources, querying, and analysis of the data, and putting the results into practice. BPM is the practice of improving the efficiency and effectiveness of any organization (see Table 6.2) and is gaining momentum as businesses become more focused on the bottom line and the cost and benefit of every function.

Developments and advancements in the past few years have made real-time enterprise-enabling technology readily available to most enterprises. However, this has led organizations to a more fundamental challenge. With

TABLE 6.2 Why BPM?

Helps businesses discover efficient use of their business units, and financial, human, and material resources by continuously assessing the present state of business and prescribing a course of action

Optimizes processes by creating better feedback loops while continuous and real-time reviews help to identify and eliminate problems before they grow

Assists enterprises, takes corrective action in time to meet projections, coming up with a plan to overcome potential problems, and for risk analysis and predicting outcomes of new initiatives such as merger and acquisition

Enables enterprises to monitor efficiency of projects and employees against operational targets, while integrating the processes with CRM or ERP the tools to get the information needed by the managers, the big question is to determine what that information is and who should be provided with this information.

An exhaustive understanding of the enterprise-specific business functions is a must for answering this question as each organization has business processes and practices that are unique to its own business model. In-depth insight about a business including what it creates, who it is trying to serve and how, where it is trying to go in the future, and how it will get there is the key to unlocking this puzzle.

The benefit of BI technology is often diminished because many customers struggle to identify and capture the most important information about their business. Furthermore, before BI technology can enable real-time connections among people, processes, and data, managers need a process for defining what these connections should be, and identify those that are time critical enough to require real-time data movement, warehousing, monitoring, reporting, and alerting. Enterprises must have guiding principles that focus their efforts and an easy-to-follow process for defining a linked network of critical success factors.

BPM modules exist within BI systems to assist managers in monitoring existing business operations and to provide guidance in identifying improvement opportunities. BI and BPM systems enable organizations to create dynamic information systems that link managers to business data, making possible the fast and accurate decision making that's required of an agile enterprise.

In order to use BPM effectively, organizations must stop focusing exclusively on data and data management, and adopt a process-oriented approach that makes no distinction between work done by a human and a computer. The idea of BPM is to bring processes, people, and information together to improve the performance of the organization.

The Real-Time Enterprise

Real-time enterprise fulfills the demand for timely and relevant information vital for making informed decisions. Informed decisions are better decisions that help the organization meet its goals and have a very profitable future (see Figure 6.3). RTE puts appropriate information at the fingertips of key decision makers at the right time while presenting information in an actionable way.

Automation offers the most efficient methods for retrieval and dissemination of this information. Unless substantial portions of human effort in this area can be replaced by automatic operations, no significant overall improvement can be achieved. However, information retrieval processes

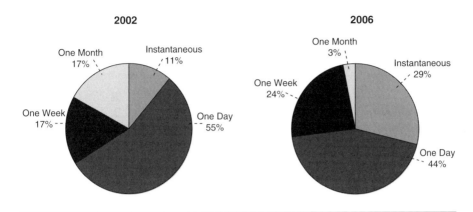

Figure 6.3   Changing data landscape: data latency

mechanized so far still require appreciable human effort to organize the information before it is entered into machines.

The real-time enterprise harnesses new technology advancement to permanently fill in the information blind spots and completely automate the broken processes that plague enterprises today.

What Exactly Is Meant by Real-Time?

Real-time may mean different things to different people. In many cases, it is fine to have a report run daily against a database that is updated once each day. Yet, at other times this might be insufficient. Some define real-time as analysis and reporting based on data entered into an operational system or database for reporting and analysis within an hour; to others, real-time means any information entered in the previous hour. But there are instances where this is not good enough.

The reality is that different applications, and even different usage requirements for the same application, necessitate different degrees of information latency. As cycles continue to shorten, more business intelligence and transactional systems will rely on a hybrid of operational and warehoused data depending on the user need. An appropriate definition of real-time is the ability to meet necessary latency requirements for an application or a user.

This leads us to another important question.

What Is a Real-Time Enterprise?

According to the Gartner Group a RTE is "an enterprise that competes by using up-to-date information to progressively remove delays to the management and execution of its critical business processes." Gartner also describes RTE as a "focus for change that necessitates the deployment and exploitation of IT from a business perspective ... it is not a technology in search of a problem."

Why Real-Time Enterprise?

The basis of the real-time enterprise is instantaneous availability of information, which calls for marrying an organization's business strategy with a computing architecture that is flexible and can be changed to meet new demands (see Figure 6.4). The real-time enterprise preserves business agility while reducing business cycle times enabling organizations to gain better insight into enterprise operational and financial performance through immediate alerts based on events.

More importantly, however, the RTE fosters the creation of a collaborative environment. The system supports various communities of interest within the organization that come together for a time to solve a cross-functional business issue and disband when the issue is resolved. It also supports seamless communication and standardized access to information for business partners, suppliers, and employees. The benefits of this collaborative environment are enormous. First, it provides users with "one-stop shopping" for information, as it presents a single gateway to organizational

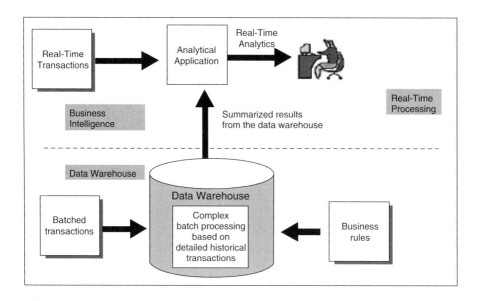

Figure 6.4   Real-time processing

information and integrates stovepiped information into shared solutions. Second, corporate information is centralized, enabling improved management and development of content (i.e., version control, tracking, and archiving).

The system also enables identification of business opportunities, monitoring programs, improved decision making, and seamless access to disparate data sources, without the need to implement a worldwide enterprise resource planning, customer relationship management (CRM), or supply chain management (SCM) system. It also enables business activity monitoring and immediate response to business events.

Managers no longer have to manage by looking through the rearview mirror at old data, but can spot problems early and take immediate action. Managers can migrate decision making from reactive to proactive. Enterprises are more motivated than ever now to set strategic targets for reducing end-to-end cycle time for the processes that are most critical to their various value disciplines and strategies with real-time enterprises seen today by CEOs and CIOs as one of the most important mechanisms for leveraging business value from enterprise IT investments. These processes may be as diverse as the routine operational day-to-day processes (such as the order-to-cash cycle) or they might be high-impact strategy areas, mergers and acquisitions, or new product development.

At most enterprises, executives receive reports and then start asking questions. They have intelligent tools to analyze problems, but they typically use them after the fact. This is a reactive way of running a business. If managers are properly equipped with information, they would know immediately when customers cancel orders and would be on the lookout for patterns and take corrective measures to stop the same rather than discover the loss once a substantial number of customers have already moved to the competitor. Perhaps these customers were confronting a common problem that led to dissatisfaction. These are issues that must be identified and dealt with before they become major problems.

Better execution can easily be termed as the main reason enterprises are adopting real-time business technology solutions leading to RTE, and organizations have identified customer support as the business function that can benefit most from real-time information. In addition to customer service, the most frequently identified driver for real-time information also is the key motivation behind investments in enterprise business intelligence solutions.

Value Proposition

The value proposition of the RTE is clear. It leverages existing IT investments and creates a collaborative, information-rich environment that gives management what it cannot get enough of: instant answers to their business questions. That is as close to a sure-fire formula for success as you can get in today's topsy-turvy world.

With the increasing complexity of today's businesses which have terabytes of data dispersed in different systems and databases all over the enterprise and the Internet making corporate data along with corporate response efficiency grow at an astounding pace, demand for RTE will continue to grow commensurate with the increase in corporate data and the need to deliver useful information to people within and beyond the enterprise in real-time.

The RTE allows enterprises to accelerate the speed of information throughout the enterprise while automating, auditing, and continuously refining key business processes. The end result is faster decision making and performance management, shortened business cycles, improved responsiveness, and an architecture that supports constant innovation and rapid change.

Enterprises that move quickly to build the RTE will enjoy significant competitive advantage and reduced costs. They will also realize an increased return on their existing investment in packaged applications, many of which have not lived up to expectations, frustrating IT and senior managers alike. By achieving where first-generation business intelligence and point-to-point integration solutions failed, the RTE will dramatically increase the strategic value of IT in an organization. RTE has five main strategic benefits.

Reduction of Waste and Inefficiency

Most new business processes are designed with advanced Internet technologies of the present era at the center and therefore take advantage of the efficiencies afforded by these technologies, thereby reducing waste and inefficiency within the organization.

Reduction of Latency and Elapsed Time

More efficient business processes mean reduced elapsed time hence latency. The fact is that a longer latency means reactive, rather than proactive, management of an organization. Most organizations are complex systems rife with latency deficiencies and executives still make key decisions based on batches of critical data that is days, weeks, or months old by the time it reaches their desks. This can be fatal in today's information-driven economy. Reduced response time also often generates significant cost savings.

Competitive Customer Service

With the advent of the Internet, customers have much more access to critical information without human intervention. This empowerment has led to higher expectations and demand for faster response times, or the customers will simply take their business elsewhere. Customers are demanding true transparency of information for making the right choice. These real-time customer interactions go both ways though: enterprises can take data gleaned from customers to offer improvements or customized product offerings on the fly.

Better Management Decisions

Managers can reduce risks and improve outcomes of decisions by making more informed judgments by relying on up-to-date information. The ever-increasing pace of business is causing managers to make decisions under greater time constraints. Managers have to gather, analyze, plan, and communicate decisions as quickly as possible and the earlier that complete and accurate information is given to them, then the better quality the decision will probably be.

More-Transparent Management Decision Making

Managers with little or low-grade information often have to base their decisions on assumptions and not on facts.

REAL-TIME INITIATIVES

Today's knowledge workers have it relatively easy when it comes to using business intelligence tools. A decade ago, users had to line up at the door of the IT department waiting for their report requests to be fulfilled. In some cases, they waited days or even weeks to receive reports. Today, users typically generate their own reports, use their dashboards, and control their embedded analytics right from their desktops through a standard Web browser or enterprise portal, or through office applications such as Microsoft Excel.

But for the IT department, which supports today's desktop analytics, coping with new information, new users, and new query techniques brought about by RTE is not so easy. The volume of data generated by business is doubling every nine months, twice as fast as Moore's Law, leading to a gap between the computing infrastructure available and required. This gap is even further exacerbated by the escalating need for near-real-time results. Adding to these volume challenges is the fact that business intelligence technology itself is becoming more sophisticated, with an increasing need for in-depth analysis as enterprises move from standard reporting toward interactive ad hoc BI and discovery analysis.

Increased sophistication laid out by RTE-enabling tools has made event-driven, real-time analysis possible and even expected by many users. In an ideal world, the IT team would be able to deliver these functions while at the same time enabling self-service capabilities and using rapid prototyping to test out newer generations of tools and applications. IT would be a change enabler, not a bottleneck. But in the real world, IT managers are squeezed by these conflicting demands: the need to deliver advanced business intelligence tools to users in order to maximize adoption while at the same time keeping costs within budget. These demands, combined with a growing number of users and a growing volume of data, present a serious challenge in achieving RTE.

Attaining RTE

The success of the RTE rests upon the fact that it uses the knowledge repositories that reside within the organization as the underpinnings of the solution. The architecture collects the data from these repositories and combines it with other pertinent information resident within the organization, tacit information that may not be in a database, but that is critical to the operation of the business. This information includes, but is not limited to, content from policies and procedures, program management, customer relations processes and interactions, published and unpublished methodologies, and training information.

Leveraging the integration of portal and BI technologies, the organizational data and other—mostly unstructured—content is coalesced to form an organizationwide integrated knowledge base. This integrated knowledge is presented via Web services to the enterprise portal, which may be accessed through organizational intranets, the Internet, and extranets. Access is based on a rights management system that manages individualized security and user protocols and IDs. Portal users can receive information on their laptops, PDAs, and wireless phones (and even their traditional clunky desktops!).

Users can then customize their information environments to see only the content they need to do their jobs. More importantly, however, they can use the variety of information touch-points available to them to participate in a truly collaborative environment to solve business problems.

The RTE can be best attained through the right application of modern enterprise business intelligence and enterprise integration technology and best built incrementally so enterprises can realize near-term results that can be easily reused for future solutions. This requires partnering with an experienced vendor that provides both enterprise business intelligence and integration technology needed to support RTE efforts today and down the road.

Building an RTE to take advantage of the benefits it offers requires organizations to avoid the trap of short-term thinking. RTE must be handled as a long-term goal. Enterprises must reengineer their processes, improve or replace major applications, and they must use business intelligence to present information to managers quickly.

Unlike the common belief, RTE is not one system; there are several key solution areas that collaborate to build the real-time enterprise. Again contrary to general perception, these are not bleeding-edge technology projects that depend on risky early adoption, nor do they rely on the deployment of a monolithic technology platform that carries a huge up-front cost. Organizations have to build the RTE infrastructure application by application, process by process, and partner by partner. They can embrace real-time solutions in a single step by creating complete solutions from scratch or can build a solution around existing IT resources incrementally, realizing value each step of the way. To do this they must redesign their software infrastructures to create an Enterprise Nervous System (ENS).

Established as well as upcoming enterprise business intelligence suites and enterprise application solutions form the cornerstone of the ENS, hence real-time enterprise. The key technologies of an ENS are:

- Business process management software that controls the flow of work between steps in a business process. These steps may be either human-assisted or automated.
- Database Management Systems (DBMS) that hold and manage the information needed by the major business applications.
- Data warehouses that contain the data and support the tools needed to provide decision makers with insight into business operations.
- Knowledge management tools to provide workers and decision makers with the knowledge and skills they need to play their parts in real-time processes.
- Enterprise portals to give staff easy access to up-to-date information along with a content management system to ensure that unstructured corporate information is up-to-date.
- Integration brokers that link applications in real-time without loss of meaning or data.
- Real-time analytics software that provides sophisticated analysis of data in real-time.

DIFFERENT FLAVORS OF RTE

Many enterprises have already implemented RTE applications and practices and are realizing very real benefits. The future should see many more enterprises beginning to take advantage of the competitive benefits offered by an RTE by choosing to make the investment to upgrade legacy systems and move key information toward the fringes of the organization where managers can use it to make informed real-time decisions.

RTE solutions fall into the following categories:

■ Enterprise business intelligence and performance management
■ Service-oriented integration and B2B collaboration
■ Event-driven integration and business activity monitoring

Enterprise Business Intelligence and Performance Management

Enterprise business intelligence is the solution for unlocking information buried in operational systems and accelerating the flow of information across the extended enterprise. Most organizations are aware of the capability of business intelligence and reporting as a means of gaining better insight and making better decisions; however, few have taken the necessary steps to realize the full potential of BI.

In order to be a key enabler of the real-time enterprise, a BI solution must become ubiquitous. In order to do this, it must provide comprehensive data access, incorporate both staged and live operational data when required, deliver appropriate levels of information to anyone in the enterprise regardless of level or role, allow users to manipulate data in reports to reach further conclusions, and scale to accommodate increasing numbers of users.

Enterprise business intelligence is also the key enabler for Corporate Performance Management (CPM). CPM allows enterprises to proactively monitor and eliminate impediments to performance before they affect financial results. Corporate performance management is a real-time initiative that aligns everyone in the organization in support of a business strategy by proactively monitoring and measuring business performance and, when problems occur, alerting key personnel at all levels in time to take corrective action.

Service-Oriented Integration and Collaboration

The next generation of integration solutions has emerged to eliminate the business process white spaces that constrain profitability and agility. Unlike the point-to-point solutions that marked the early era of integration, service-oriented integration maintains the flexibility required for continuous growth. This is why Service-Oriented Architectures are so attractive.

Service-Oriented Architectures (SOAs) are rapidly gaining importance with every passing day, becoming the buzzword among enterprise technologists. From a business perspective, a service-oriented architecture affords an organization the flexibility to adapt as the business grows and as processes mature and evolve. At a technical level, an SOA provides a standard programming model that allows components to be published, discovered, and invoked over a network. This model provides the infrastructure for an organization's systems to be agile in responding to changing business conditions, yet allows technical implementation details to remain transparent.

SOAs employ open standards that allow development teams to rapidly assemble new applications by calling existing applications "services," maximizing the value of existing systems, raising the level of abstraction to the business-process level, shortening deployment times, and reducing risk by limiting the amount of code required to build new applications.

The service-oriented architecture's flexible design means that its growth path is essentially unlimited. The bottom line is that service-oriented architectures allow enterprises to get projects into production quickly and without investing millions of dollars up front, or giving up the flexibility that they need to keep their businesses agile and profitable.

The maturation of Internet standards, including J2EE, JCA, XML, .NET, SOAP, and WSDL, makes the deployment of service-oriented applications that connect people and machines to automate business transactions possible. Furthermore, it is these standards that make the entire concept of a service possible and make these applications resilient to change.

Benefits of SOA

An SOA provides a range of both business and technical benefits for the corporation.

Business Benefits

■ *Leverage existing investments:* Organizations have spent time and money developing their existing infrastructure. Today's dynamic environment demands new uses of systems and process information to help enable more efficient processing and cost-effective business operations. Repurposing systems, processes, and data from existing systems provides one avenue for managing the cost. Exposing legacy systems as services also creates an environment that shares information and processes from systems that were previously disparate or connected through point solutions.

■ *Faster time to market:* An SOA inherently promotes reuse as part of its core philosophical approach to development and integration. Reuse of services and components allows new applications to be quickly assembled to respond to changing market conditions or business demand.

■ *Risk mitigation:* SOA-based efforts reuse core services and processes that have been developed and tested and are well understood, increasing the level of project success of an implementation by reducing potential bug introduction. In addition, once the core architecture has been implemented, development occurs at two levels. One level focuses on orchestration, the assembling of services into processes, workflows, or applications, which requires a focused skill set that is more business oriented. A second level involves the design and development of the services and underlying infrastructure. This separation of skills improves risk management from an organizational perspective by allowing effective allocation of resources to efficiently deliver development tasks.

■ *Continuous improvement:* Service- and component-based development provides continuous improvement because systems and processes communicate via services and interfaces, the implementation of which is encapsulated and invisible to the requestor. Hiding the underlying implementation provides an opportunity for continually improving and optimizing the underlying code base without affecting the use of the service.

Technical Benefits

■ *Location transparency:* Decoupling the client requesting a service and the service itself is known as location transparency. The SOA allows a request to not know (or care) where a component or service is located because of the publication and discovery mechanisms. This eliminates locating, understanding, configuring, and incorporating remote functionality from a development standpoint. The SOA framework should handle all of the publication and discovery work required to use a service. Location transparency is facilitated in an SOA by support for standards such as UDDI in the Web services world.

■ *Loosely coupled:* Coupling refers to the dependencies that exist between software components. When discussing coupling, two main definitions are covered, tight coupling and loose coupling. Loosely coupled implies that decisions can be made by services, components, and applications at runtime rather than compile-time. Loosely coupled components can act independently of each other, whereas a tightly coupled approach requires components and dependent components to be available for binding at compile-time as well as runtime. An SOA promotes the design and implementation of loosely coupled applications or platforms by providing mechanisms for supporting loose coupling, and by ensuring that every service is decoupled in time, protocol, and location.

■ *Late binding:* Late binding allows a loosely coupled application to be flexible because there is no inherent knowledge of how the application will be used. It refers to components that determine behaviors and relationships at runtime, rather than at compile- or deploy-time. This dynamism allows applications the flexibility to adjust to changing requests and responses while interactions are occurring. This late binding paradigm allows systems to be reusable, extensible, self-assembling, self-healing, and more maintainable.

■ *Protocol and device independence:* Services are defined in a manner that is independent of device (either for distribution or presentation), connection mechanism, or transportation protocol. This independence provides the SOA an opportunity to present information to respond to a request from mobile users, desktop users, applications with high or low bandwidth, legacy systems, and so on. Allowing a network to perform in a heterogeneous fashion lets a service "be serviced" by the appropriate system regardless of the network on which it resides.

SOA Types

Two major types of applications comprise service-oriented integration: composite applications and multistep processing.

Composite Applications

Composite applications provide new functionality while incorporating transactions from existing packaged applications or legacy systems. They are often built to deliver a single new user interface that eliminates the need for users to interact with multiple systems separately, but they may also perform application bridging to let several applications share related business logic with each other.

Composite applications function by creating new business logic on the middle tier that calls existing systems to perform a specific function that spans multiple functional areas. Unlike point-to-point integrations, they are loosely coupled and isolate the service-enabled system from change. This makes them easy to evolve over time.

Multistep Processing

Multistep processing, more commonly called Straight-Through Processing (STP), manages a larger-scale integrated business process that uses many physically independent applications. Unlike composite applications that focus more on user interfaces or application bridging, STP emphasizes the interaction of highly complex processes.

An organization can build the RTE incrementally by deploying a series of existing composite business applications and then using an integration broker to build straight-through processes by orchestrating the composite applications. This exemplifies the "system of systems" that is one of the hallmarks of the real-time enterprise.

By delivering a series of composite applications that provide process improvements and value and then reusing them for STP applications that close off the white spaces completely, enterprises can enjoy near-term results that can be used to fund higher-value straight-through processes. This also tends to mitigate the risk of building an STP application from the ground up.

Service-oriented applications can be easily extended to customers and partners by using an existing EDI connection or by making the application available as a Web service. Many vendors are now offering Web service solutions that allow BI content to be integrated with other information, or fed into other applications. This creates exciting opportunities for building BI into the service-oriented architecture, and conversely for using the service-oriented architecture to short-circuit the labor of traditional BI solutions such as data warehousing. Although the Web services model is gaining momentum, the scenario, best described as an edge-integration strategy, has the benefit of reusing existing EDI connections where they exist, while leveraging the security advantages of EDI and Internet EDI.

Web Services for Business Intelligence

Business intelligence, an important management function, has undoubtedly benefited from a range of technological innovations. The business intelligence space is being radically challenged by new forms of computing including Web services and grid computing; meanwhile, the use of the Internet as a platform for business intelligence is becoming more mature and sophisticated. There is an important role for Web services in the business intelligence space; taking the cue, some of the specialist BI vendors are starting to support Web services.

As both concepts, BI as well as Web services, are fast gaining acceptance, the convergence of BI and Web services will become mainstream in a fairly short timeframe, possibly even earlier than the use of Web services for pure transaction processing. To date, the deployment of Web services technology for commercial transaction processing has been limited. However, business intelligence already uses many external sources of information, and may therefore be more receptive to the use of external Web services where this provides economic or technical advantages. Web services are being widely discussed for integration of operational business processes but will be as effective for management systems including BI, and should not be restricted to operational systems.

Many enterprises are starting to deploy Web service technology for connecting applications internally. There is significant interest in deploying the same technologies externally, for connecting applications between multiple organizations, although this is currently inhibited by concerns about security and the immaturity of adequate standards. However, the deployment of Web service technologies across the organizational boundaries and the construction of a services-oriented architecture spanning multiple enterprises will be soon embraced by organizations.

Event-Driven Integration and Business Activity Monitoring

Event-driven integration is a special type of real-time initiative that spans multiple systems across the enterprise. Event-driven integration calls for creating applications that "listen" for business events, process them, and execute processes to resolve problems or recommend process improvements based on those events.

Like service-oriented applications, event-driven applications are loosely coupled, modular, and fully encapsulated, which means that they do not have to have an established relationship with any system they communicate with and can be changed without affecting any system in the enterprise or beyond. However, event-driven processes don't rely on a request-and-response relationship between applications, but rather a listen-and-push metaphor.

Most service-oriented applications occasionally "poll" for new requests, run a process, and then issue a response, however, event-based applications are always on, intelligently absorbing events, processing them, and immediately sending real-time messages to systems and people when a particular set of conditions is met. It is this proactive "pushing" of real-time information and messages that distinguishes event-driven integration. Event-driven integration is the ultimate realization of the "sense-and-respond" business environment that defines the real-time enterprise.

Business Activity Monitoring (BAM)

Business Activity Monitoring (BAM) is an advanced instance of event-driven integration that is enjoying increased adoption. Business activity monitoring is bringing business one step closer to the goal of the real-time enterprise.

BAM deploys operational business intelligence and application integration technologies to continually refine automated processes based on feedback that comes directly from knowledge of operational events. In addition to auditing business processes (and business process management systems) and sending event-driven alerts that trigger process adjustments, BAM solutions also can be used to alert individuals to changes in the business that may require action. And BAM data points can provide aggregated insight to executives who are planning strategically.

Enterprises and government agencies alike are already implementing BAM solutions and continue to refine those solutions as they approach the joint goals of zero-information latency and self-correcting processes.

BAM Building Blocks

BAM has three main components:

1. *Event absorption:* Event absorption refers to the collection of events from multiple applications as they occur. ETL (Extract, Transform, and Load) tools help provide access and transformation of these events into a usable data model, where they can be analyzed with historical contextual data from a data warehouse.
2. *Event processing and filtering:* Event processing and filtering occurs when real-time events are analyzed in context and rules are applied to determine if there is an irregularity that needs to be reported. The application of these rules often involves a multistep process that generates its own data.
3. *Event, action, delivery, and display:* Event, action, delivery, and display describes how the system handles alerting various key decision makers when there is a problem. This entails generating reports based on the information generated in the BAM process.

Because BAM combines both enterprise business intelligence and enterprise integration solutions, all of the same barriers that apply to the successful implementation of those technologies also apply to BAM.

BARRIERS TO REAL-TIME ENTERPRISE

Most BI implementations are plagued by barriers that limit the strategic value of BI as a key component of the real-time enterprise. Most organizations have experienced the following pitfalls.

Absence of Enterprise Standard

There is no enterprise standard for business intelligence. With multiple departments running disparate tools on a local level, it is impossible to get answers to bigger questions or empower real-time decision making by everyone in an organization and its extended enterprise. Not only do these redundant tools cost enterprises money, but also the answers to important operational questions are confined to a functional area or line of business.

Client/Server Architectures Rather than the Web

Almost all of these tools are based on client/server architectures rather than the Web. This complicates the delivery of information, by requiring a client installation for someone to receive reports or perform further analysis, which limits the usability of the product to a small minority: addressing the information needs of the few at the expense of the many. Even if the information is both timely and accurate, it has limited value if it has a limited user base.

The real-time enterprise by definition is inclusive, and the inability to deliver intuitive reports via a browser makes sharing information beyond the organization's four walls a virtual impossibility. To support information delivery on a broad scale, enterprises must standardize on an Internet-based product that allows users to access reports directly from their browsers.

Exclusive Reliance on Data Warehouse

Data access and the ability to support scheduled and ad hoc reports from both transactional and reporting databases have serious implications for data architecture and management. Data access is fairly straightforward but its importance cannot be overlooked. Without complete access, the information consumers (both end users and applications) will get an erroneous picture. For years, conventional business intelligence and data warehousing wisdom held that business intelligence applications should access information exclusively from a data warehouse to avoid putting a drain on operational systems.

This, combined with the fact that most BI tools were optimized to report from a data warehouse and could not report directly from operational systems, has meant that enterprises have undertaken massive data warehousing projects at a huge price and considerable risk, in order to have a suitable data infrastructure for business intelligence and reporting. Even with ETL tools to automate the movement of the data to the appropriate warehouse, operational data store, or data mart, project cycles and solution paybacks have proven lengthy.

The problem with this, in addition to its huge cost, is the amount of latency it creates. There are times when real-time does mean instantaneous, especially when someone is trying to pinpoint a problem. Even if the data warehouse and application reports against it get updated on the hour, it's not good enough. As business cycles continue to shorten, and more businesses adopt real-time models, these situations will become more frequent. Without the ability to access live data, it is impossible to know whether you have the right answer. This could be the difference between making earnings estimates and receiving a warning.

Scalability Constraints

Again, BI products with client/server architectures cannot grow to accommodate thousands and hundreds of thousands of users. As enterprises embrace the real-time enterprise, they will roll out applications that will deliver information to an ever-growing volume of users. If the BI application cannot handle this, or requires prohibitive hardware spending, it is safe to assume that it is not the right tool for the RTE.

7

TELECOMMUNICATIONS INDUSTRY: CHANGING LANDSCAPE

The telecommunications industry expanded substantially in the past decade. Technology advancement along with the liberalization of once closed markets and privatization of government-held monopolies changed the nature of the industry in the 1990s and continues to shake up the industry every now and then. In early 2000, the industry scaled new highs with respect to market capitalization. Both business and technology disruptions have introduced significant expansion and innovation (see Figure 7.1).

Since then, however, the telecommunications industry has been overwhelmed by a series of events that have led to a depressed industry segment. Over the past few years, it has suffered from severe debt, overcapacity, customer churn, and service commoditization. The earnings declines and flight of capital have driven the industry to reevaluate and reassess fundamental business practices and devise survival strategies that will lead it back to competitiveness and consistent profitability.

The same forces that fed the development of new services and the entrance of new players also saw margins grow slimmer for most services as well as significant customer churn as competitors offered alternative choices. The expansion of the infrastructure has not been absorbed yet with an equivalent rise in demand and profitability. On the other hand, investments incurred for expansion and for acquisition of 3G licenses are becoming a back-breaking exercise for the once solvent and profitable communication service operators. Trillions of dollars in market capitalization have evaporated in the past couple of years and hundreds of thousands of jobs have disappeared from the sector. The telecom industry, which is

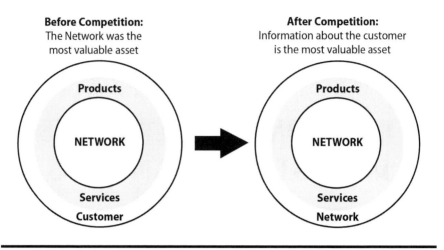

Figure 7.1   Telecom changing landscape

undergoing a period of difficult change, is under a great deal of competitive and market pressure. The major challenges facing the industry are:

- Increased customer dissatisfaction with existing telecom services
- Market uncertainty and excessive debt
- Bandwidth commoditization
- Limited market capital
- Large, expensive, and inflexible IT infrastructures

In today's extremely challenging business environment, many telecommunications operators and carriers are measuring their success by the size and growth of their profit margins. As a result, carriers are under intense pressure to reduce or eliminate the major threats to these slim margins including revenue leakages and frauds, inaccurate or missed intercarrier billing, churn, inefficient network usage, and least-cost routing plans. These competitive and market pressures are also making the telecom industry reassess its business model and redefining the path that will return it to competitiveness and profitability.

4 Cs OF THE TELECOMMUNICATIONS INDUSTRY

There are four key challenges faced by the telecommunications industry today, described as the 4Cs. As with financial services, deregulation is driving many of these currents as a metadynamic that influences all of these trends:

- Consolidation
- Competition

- Commoditization
- Customer service

Since deregulation, the ability to compete in a much wider array of markets has opened established telecommunications companies, and prospects of new entrants gaining market share have also become a reality. This has resulted in existing telecommunications companies buying the capabilities to enter new markets and new players gaining substantial market shares. This has resulted in tremendous merger and acquisition activity along with increased competition. As a result of these trends and regulatory pressure, there is a significant emphasis on customer service that did not exist before.

TOWARD A CUSTOMER-CENTRIC BUSINESS MODEL

In competitive telecommunications environments, customers choose their service providers. Today, this is a reality for all sizes of telecommunications companies as well as all types of telecommunications companies, whether long-distance, Internet Service Provider (ISP), wireless, or local POTS.

Under competitive conditions, the customer becomes the central focus of the carrier's activities. Customer requirements not only determine service offerings, but also shape the network and affect the organizational structure of the carrier's focus on particular types of customers. With the customer at the center of the telecom enterprise strategy, key to survival for these providers and operators today is to focus on the very basics of business such as retaining and servicing existing customers as well as reducing the cost of operations. Nearly all telecommunications companies today are responding to a mission-critical need to compete more effectively, as a result of:

- Rapidly changing, increasingly competitive, and global markets
- Increasingly volatile consumer and market behavior
- Rapidly shortening product life cycles

To do so, it is necessary to analyze accurate and timely information about operations, customers, and products using familiar business terms, in order to gain analytical insight into business problems and opportunities.

AT THE CROSSROADS

The business landscape of the telecommunications industry is quickly evolving. The previous model, shaped by a handful of competitors in each country, is being replaced by a model shaped by hundreds of competitors vying for a global presence. To survive in this environment, telecommunications companies can continue marketing their products to the masses. This market-share strategy has been very popular in the telecommunications industry in the past. To compete, companies are driven to increase advertising and marketing costs aggressively while discounting their products. Unfortunately, this strategy has driven customer loyalty to an all-time low. For example, it is not unusual for a consumer to switch service providers twice in the period of a single year.

This may be why companies can report a 40 percent disconnecting rate over the period of a year and still show an increase in market share. Clearly, this model for doing business presents significant challenges and over time threatens to drive profit margins unacceptably low. The ultimate evolution may be similar to what has been seen in the retail industry, where a good year produces profit margins in the two to three percent range.

Companies can focus on tailoring products to the individual customer. In this "share of customer" environment, customers are differentiated in addition to products. Corporate resources are efficiently allocated to customer care in relation to the customer's lifetime value. Those customers whose loyalty can be earned and whose lifetime value to the company is high will receive a majority of the attention. In contrast, customers who are not loyal or whose lifetime value is low will receive a lower degree of attention. The result will be an environment that optimizes profits by nurturing valued customer relationships. Essential to this strategy will be the ability to leverage evolving technologies to accomplish the following:

- Understand the customers' needs and behaviors.
- Leverage this understanding to identify, develop, and deliver relevant products and services.

Choosing or integrating these strategies and migrating to this new environment is one of the most profound decisions facing telecommunications companies across the globe.

THE CUSTOMER IS KING

As telecommunications markets become increasingly competitive, the ability to react quickly and decisively to market trends and to tailor products and services to individual customers is more critical than ever. Although data volumes continue to increase at an astounding rate, the problem is no longer simply one of quantity.

At the heart of the issue is how companies are using their information. Increasingly, particularly in the telecommunications industry, it is important to understand customer preferences and behaviors. It is imperative to understand all the parameters of a customer, whether individual or otherwise.

Although this sounds simple enough, a telecommunications company faces several hurdles in achieving this objective and in targeting its product lines to current or prospective customers. Ironically, the telecommunications industry is in a unique position to understand the customer because it can direct its energies in various channels to obtain customer information.

Strategic Shift—From Product to Customer

Many companies are aggressively moving (or have already moved) from a business model based on a product strategy to a business model based on a customer strategy. This environment is characterized by customer relationships, product customization, and profitability, and is in response to pressures transforming the business landscape throughout the telecommunications industry.

Growing Consumer Demand

Customers (or consumers) are expecting companies to understand and respect their needs and desires. In this world, the customer drives the relationship. It is the role of the business to hear what the customer has to say and respond by delivering relevant products and services (what they want) on their terms (how they want it). Companies can no longer expect to sell several products and services to the masses (mass marketing), but must tailor many products and services (i.e., mass customization) to the individual. This is generally referred to as mass customization.

Growing Competition

The ability to refocus a product mix in response to evolving competition is a critical success factor for any business. The key is to be able to anticipate the needs of the marketplace before one's competitors. It is this ability to outpace competitors that most companies find difficult or impossible to do, given today's amalgamation of technologies and architectures. Why is this important? Corporations today are facing more and more deregulation; mergers and acquisitions are blurring the relationships to customers; and globalization of the marketplace and consumer is opening up businesses to new avenues for expansion and, as a result, new competitors. Therefore, it is mandatory for a corporation to restructure itself quickly without losing the ability to compete.

Optimization

The ability to measure and predict Return On Investment (ROI) is something that corporations find difficult to perform rapidly. These measurements indicate the health of the corporation, and the ability to determine them rapidly allows a corporation to change its direction with minimal

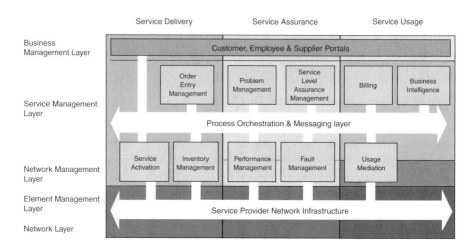

Figure 7.2   Telecom enterprise architecture

loss. Other examples of the need for optimization include the ability to determine the most efficient channels for contacting customers, target the appropriate customers for a corporation's product mix, and identify new product opportunities before the competition.

Technology Constraints

Unfortunately, most information systems at telecommunications companies are built around the product strategy business model. This has resulted in a variety of product-oriented systems that effectively run day-to-day operations. Carriers struggle with their existing patchworks of general-purpose data warehouse solutions to store and analyze the mountains of data they create every day. Large networks and their associated switches, billing systems, and service departments can generate hundreds of millions of terabytes daily. These terabytes of dynamic customer data will continue to grow exponentially as carriers add new services and as IP-based traffic increases. This ever-expanding volume of data puts a strain on the performance capabilities of today's traditional relational databases, servers, and storage systems that provide the foundation for Business Intelligence (BI; see Figure 7.2).

Customer demand for new services, such as third-generation wireless networks, consolidated billing, and consistent and reliable service, has been affected by technology limitations. The difficulties created by legacy back-office systems, known as Operational Support Systems/Business Support Systems (OSS/BSS), are primarily rooted in their complexity, scale, rigid operational requirements, lack of interoperability, and lack of service focus. This has led to enormous challenges when attempting to deploy new services and adapt to rapidly changing customer needs.

In the past decade, the service providers have spent significant resources and energy installing systems for operations management and business process automation both for business and operational support. However, the complex questions that need to be answered go beyond any one operational system. Today telecommunications companies might know who their customers are, and are marginally effective in marketing new services to them. However, few of them are equipped to know who their profitable customers are, which services these cream customers use that make them profitable, and which marketing campaigns can be targeted to this segment.

Most of the rudimentary systems available or used in the past by these telecommunications companies are those where they find themselves having to extract data from multiple systems and manipulate complex spreadsheets. This exercise is time-consuming and fraught with delays. If any parameter changes in the equation, they find themselves at square one, without the ability to construct what-if scenarios and respond effectively to the changing market conditions.

TECHNOLOGY CAN HELP TELECOMMUNICATIONS COMPANIES

Carriers rely on analyses of their terabytes of customer, product, and traffic data to help them make business-critical decisions that will positively affect their bottom line. High-end data warehouses and powerful business intelligence solutions are essential tools to help carriers meet profit goals. Analyzing and integrating in-depth data from multiple departments enable carriers to reduce revenue leakage and churn, mitigate fraud, optimize network usage, and increase profits.

Competition makes things more difficult for operators, and the analysis they need for operations, to stay ahead, needs to be much more granular to allow for true assessment of the profit contribution of the customer and the services, especially when most of the business questions they want an answer for go across multiple operational systems as business processes flow across many departments.

Telecom companies need to deploy new and more focused business services and create stronger customer relationships. This implies a high degree of responsiveness to customer needs and concerns. Telecom companies are, however, strapped with large legacy IT infrastructures that are extremely complex to manage. These infrastructures have rigid operational requirements and resist incremental change, thereby affecting the agility and responsiveness of the telecom companies to customer demands and the deployment of new and improved services.

Telecom companies are looking for ways to build IT flexibility and agility in their businesses, in an effort to lower the costs and risks involved in upgrading and evolving these OSS/BSS. The need to be flexible and nimble requires telecom companies to migrate to more open technology environments and to adopt an architectural approach to integration challenges.

The need to deploy new services requires these companies to adopt a new generation of standards-based middleware that enables rapid and cost-effective service-oriented integration of OSS/BSS infrastructures. The new technology needs to adopt an incremental approach to the creation of service-oriented OSS/BSS architectures, thereby migrating to the IT base of telecom companies through phases of renovation and service consolidation while driving customer-focused service innovation.

Many telecommunications companies are organized first by customer segment—be it consumer or business—then by product or service, such as consumer long distance, business, or local. Managing data within these silos and across these silos can be a challenge, and data warehousing is an essential lynchpin in maintaining a profitable business model.

A robust BI solution can help telecommunications companies manage the complexities involved in this calculation. Telecommunications companies worldwide are exploring business intelligence solutions to achieve competitive advantage. The key solutions for which telecommunications companies are looking involve marketing, such as customer retention, target marketing, and campaign management, customer-relationship management, and network business intelligence, to streamline network assets. Moving forward, additional systems are needed with capabilities to deliver best-of-breed business intelligence and business management. These capabilities enable companies to accomplish the following:

■ Understand the needs of their business (business intelligence)
■ Manage actions based on those needs (business management)
■ Effectively run day-to-day operations (business operations)

These capabilities will enable companies to realize the opportunity of a business landscape characterized by customer relationships, customized product delivery, and opportunity-driven profit. One of the key enabling technologies to this evolution is the data warehouse.

BI IN THE TELECOMMUNICATIONS INDUSTRY

Telecom was among the first industry verticals to experience the benefits that BI brings to the corporate table (see Figure 7.3). Telecom was also the first to experiment with how BI, or rather analytical capabilities in conjunction with Customer Relationship Management (CRM) solutions, can improve customer experience and thereby the business. Among the first applications in this area were the telecom industry's BI initiatives to reduce customer churn. The use of BI connected to operational CRM systems helped identify

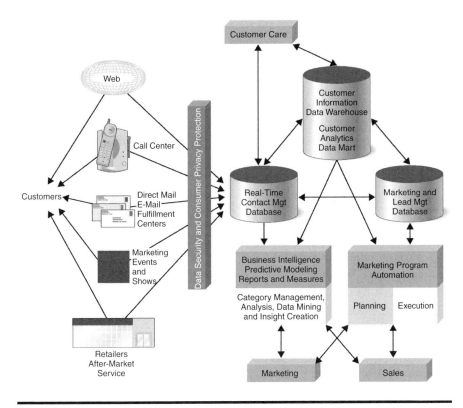

Figure 7.3   Strategic BI architecture

customers who were most likely to shift to another service provider, by analyzing the number and nature of grievances registered by users.

Although the success of these initiatives has resulted in CRM products with analytical capabilities, the case is still strong for a dedicated BI system connected to an operational CRM system, provided the linking is done optimally. This is because the new CRM products still do not match up to a full-fledged BI system's analytical capabilities, not so far at least. Today telecommunications companies can provide basic answers. However, the complex questions they need to answer today go beyond any one operational system.

For example, today they know who their customers are and are marginally effective in marketing new services to them. However, if they want to know who their profitable customers are, which services the customers use that make them profitable, and which marketing campaigns should be targeted to this segment, they find themselves having to extract data from multiple systems and manipulate complex spreadsheets.

This exercise is time-consuming and fraught with delays. If any parameter changes in the equation, they find themselves at square one, without the ability to construct what-if scenarios and respond effectively to the changing market conditions. The analysis needs to be much more granular to allow for true assessment of the profit contribution of the customer and the services. Most of the business questions they want to answer go across multiple operational systems as business processes flow across many departments.

The same holds true for many of the operational processes. For example, one area that has severe implications for customer satisfaction is the time it takes to activate a new customer. The activation of a new mobile customer can take several hours and for a new broadband customer it can take several days.

Activating a new customer touches many departments from customer and service support to accounting, credit approval, billing, network planning, network support, inventory management, scheduling a service person to configure and install the customer premises equipment, and turning the service on. Most operators have disparate systems managing the provisioning and customer activation process, and there is very little visibility of the bottlenecks and disconnects. The activation of new customers has revenue implications and, therefore, affects the share price and market evaluation of the carrier. This is a key business performance parameter with visibility in the office of the CEO and to the board.

The same disconnects happen in many other areas as well: interconnect billing, network operations and management, service creation, and introduction of new services, determining the profitability of various products, services, and rate plans. The systems at most service operators have been built for accounting purposes and do not provide adequate information across processes and functional areas to support business users in making business decisions based on the power of information.

There are software solutions that are helping service providers bridge this gap. The business intelligence applications extract and connect disjointed systems and data from disparate sources and enable the business user and decision maker to make an informed decision.

Generally, the IT organization at a service provider will consolidate data from multiple sources into large data warehouses or smaller data marts. The analytical applications and reporting tools are utilized to query the database and provide regularly scheduled production reports to the business users. Solutions have very easy to use user interfaces so the users can also create their own reports and make ad hoc queries as required by business conditions. The analytical applications and digital dashboards go a step further and let users keep their fingers on the pulse with prestructured guided analysis and sophisticated event-management engines. The user can set alerts based on predetermined business rules and thresholds.

It is important for service operators to be able to segment their customers more precisely than they have in the past. Not all services are attractive to all customers, and it is very costly to run outbound marketing campaigns that result in low response rates. Not only do they tax the resources of already overburdened call centers, but they also result in customer irritation or often times acquisition of customers who do not add any value to the bottom line. Service operators are using business intelligence tools to obtain a better understanding of their customers.

Who are their high-value customers? How can they treat them differently?

This is the same strategy that some other industries, such as airlines and credit cards have been using for years. Understand the impact of combining one service with another. Understand the basket mix of services being purchased by premier versus low-usage customers and the revenue impact of each service. Understand the impact of marketing campaigns sooner and be able to take corrective action in midstream. Without business intelligence applications, they may not know the results of a marketing campaign for three to four months. A product promotion that is run in July will have customer usage in August show up in the September billing cycle. A 30- to 60-day delay removes the ability of the product manager to take action during the campaign. With business intelligence tools the product manager can see the customer activation and usage on a daily basis without having to wait three months for the billing records. This enables her to adjust the message and the product if the campaign is not meeting its targets.

There is no quick fix out of this market. It is imperative that the service providers dive into their plentiful data and really analyze the operational and financial parameters. To obtain the margin improvements they are seeking, the carriers will have to perform detailed analyses of cost of service, customer profitability, and product margins in order to survive and thrive.

BI Requirements of the Telecommunications Industry

Telecommunications service providers should pursue an integrated information strategy from a single platform without compromising security, user requirements, or future flexibility. There are several strategic benefits to this type of "define once, deploy everywhere" solution. This type of solution enables detailed analyses within business units while allowing cross-divisional analyses by enterprise users. Once the enterprise data warehouse is built, organizations can then define data marts that meet multiple business unit needs.

Further solutions should include a multilayer security model that authenticates and authorizes access to data, protects data transmission, and controls application functionality by associated privileges. This ensures that users see only the data intended for them as access is defined for each user against project, row, and object.

Finally, the solution's integrated and flexible platform allows enterprise analysts to survey the entire enterprise to measure ROI, cost of capital, IRR, and additional financial analyses across units, campaigns, and customer segments (see Table 7.1).

Table 7.1 Telecom BI Applications

Fraud Management	Financial Analysis	Marketing Analysis
Fraud detection tool that helps management stop crime and operate efficiently. Drilling down into customer and employee contact records, it delivers insight that can reveal possible fraudulent activity, as well as identifies operational problems that can be fixed. Covers areas including: ■ Fraud analysis ■ Corrective action and notification ■ Product affinity/ bundling ■ Pricing models ■ Discounting ■ Call volumes ■ Call times ■ Response times ■ Complaint logs ■ Employee productivity ■ Capacity forecasting	This vital BI tool enables telecommunications carriers to take the financial pulse of their business whenever needed. Examination of financial performance metrics from across the enterprise arms financial managers with intelligence to make the most profitable business decisions possible. Financial insight ultimately improves gross margins and bottom line performance. Covers areas including: ■ Revenue reporting ■ P&L reporting ■ Cost analysis ■ Margin analysis ■ Tariffs ■ Taxes ■ Budget variance analysis ■ Access and line charges ■ AR/AP reporting ■ Collections reporting ■ Contracts reporting	This analytical tool makes effective category management of telecommunications services possible by providing analytics across a wide range of marketing, planning, pricing, operations, and network variables, helping management determine what promotions and service plans are most effective for specific customer profiles. Covers areas including: ■ Up-sell analysis ■ Loyalty programs ■ Customer segmentation ■ Demographic analysis ■ Cross-sell analysis ■ Service history ■ Channel efficiency ■ Next to buy ■ Promo lift ■ Price points ■ Market share
Network Optimization	Sales Analysis and Billing	Customer Care and Analytical CRM
Growing and maintaining profit margins requires optimum network efficiency. Powerful analytics tool that allows carriers to compare a wide range of metrics across	A vital tool to gain effective insight from the terabytes of data associated with selling and billing for residential, business, bundled, and unbundled services. Leverage data	Fierce competition for customers across the telecommunications landscape demands advanced customer care efforts. This BI application enables telcos to segment

Table 7.1   Telecom BI Applications

Network Optimization	Sales Analysis and Billing	Customer Care and Analytical CRM
network operations, and create real-time reports that identify problems for immediate attention. Alerts can also be created for instant notification of emergency situations requiring rapid response. Covers areas including:	analysis into competitive advantage by revealing more profitable sales opportunities and the path to more efficient back-office operations. Covers areas including:	customers by demographic, service plans, billing, and other criteria, delivering insight where it is needed, enabling managers to develop effective strategies that win and retain profitable customers while weeding out unprofitable ones. Covers areas including:
■ Traffic analysis ■ Network planning ■ Quality of service ■ Network utilization ■ Switch operations ■ Call routing ■ Capacity ■ Switch utilization ■ Volume management ■ Failure notification ■ Capacity analysis	■ Product sales and trends ■ Customer trends ■ Sales force performance ■ Commission reporting ■ Product affinity ■ Account balances ■ Utilization ■ Fraud ■ Telemarketing ■ EBPP/intelligent billing ■ Quota attainment	■ Customer scorecards ■ Churn analysis ■ Customer profitability ■ Customer plan migration ■ Service level agreement ■ Trouble ticket ■ Service complaints ■ Customer inquiry ■ Dispatch request ■ Service call monitoring ■ Preferences and permissions

BI Application Areas

A telecommunications company can use various BI tools for strategic as well as operational decision making. Furthermore, it can carry out various analyses to suit its unique requirements and position within the industry. Among the applications that play important roles in telecommunications companies' success are strategic decision support, scoring and segmentation, campaign assignment and management, traffic analysis, customer relationship analysis, corporate performance monitoring, and, last but not least, financial analysis. Other than these central application areas, other areas key to telecommunications companies' strategy are risk analysis, fraud detection (or revenue assurance), and platform convergence.

Integrated Customer View

This is the first hurdle that any telecommunications company meets. It is important to have all customer/account/transaction data at one place and to be able to correlate customers across their product holdings. In all probability, every product line within a telecommunications company sits on a separate system. The Customer Information System (CIS) should be able to correlate all customers from various product systems and define the various relationships.

If the telecommunications company does not have a CIS system, then individual customers and accounts need to be gathered from various systems and a customer integration process begun. This process can be carried out using industry-standard tools. Individual customers can be parsed using name, date of birth, address, gender, social security number, and a distinct customer ID can be assigned to a customer across all his or her product holdings.

Householding too can be applied based on specific business requirements. The telecommunications company can identify that two customers listed in both the savings and mortgage systems are actually one customer. This provides the most crucial information for a telecommunications company upon which accurate and effective analyses can be run. The data warehouse will then provide a 360-degree view of a customer. Although this process is extremely cumbersome, it is critical for effective analysis and should be treated as the starting point.

Customer Life Cycle

Every telecommunications company develops a customer life-cycle model and maps its products to this model. This methodology reflects the thinking that customers require different products and services at various points in their lives. For example, a customer using basic services may opt for mobile services and then DSL services in the future. Using this hypothesis, it is possible to segment customers in the data warehouse based on their demographics: age, location, annual income, occupation, usage, and the like. Such analysis is primarily based on a customer's position within the life cycle and the products that are applicable to that stage. This model is one of the critical factors that determine the telecommunications company's marketing campaign efforts. Every telecommunications company must have such a model and efforts should be made to integrate this model within the data warehouse analytical efforts.

Customer Financial Portfolio

Customer data can be enhanced by purchasing data from credit bureaus. Although only credit-related products are available from credit bureaus, it does provide the telecommunications company with crucial details on a customer's credit worthiness.

Customer Profitability

This is a standard analysis that any telecommunications company can carry out using a data warehouse. Analysis of a customer's profitability to the telecommunications company is crucial for campaign effectiveness. Customer profitability can be calculated in different ways. Whichever method is chosen, it should incorporate the telecommunications company's transfer-pricing mechanism at the account level. The accuracy of the profitability numbers is dependent on one factor: where will these numbers be used? If it is only for determining the type of products to be marketed to a customer, then the profitability numbers can incorporate some level of acceptable deviations.

Profitability should ideally be at the contribution level, as this number defines the net revenues that are a direct reflection of the customer's transaction behavior. Fixed costs, although allocable, are dependent on the allocation methods and can impinge on a customer's profitability.

Diversification Indicator

This strategy should accompany the customer profitability analysis. Customer profitability by itself does not signify much. First, the profitability numbers need to be compared with those of other customers. Assuming a customer's profitability is high, the telecommunications company must make the decision to either up-sell or cross-sell its products. The diversification indictor specifies the diversity of a customer's product portfolio.

The basic premise is that a customer with stable but average profitability is preferable in the long run to a customer who has high profitability but carries a less stable profitability. This indicator will have to be constructed using the product lines under which a customer holds specific products. A string field can be used to depict such holdings and a model needs to be created that assigns a specific ranking based on product-holding combinations. The ranking is based on the profitability that each distinct combination will accrue to the telecommunications company.

Product Profitability

This can be calculated as an extension to the customer profitability exercise. The idea is to determine the profitability of various products offered by the telecommunications company and to make product decisions based on such profitability. The process followed is very similar to that used for customer profitability. Again, specific focus should be placed on direct revenues and expenses. Fixed costs are important here (unlike with customer profitability) as a telecommunications company's ability to market its products is based significantly on its infrastructure, which is not a direct product expense.

Channel Profitability

This is another important aspect of the profitability exercise. A telecommunications company needs to determine which of its delivery channels are more profitable or cost-effective and should try to move its customers to the more profitable channels.

Channel profitability is a difficult exercise as most fixed expenses are usually not directly allocable to a channel. Another problem is the cost of campaigns; most campaigns go out through direct mail, branches, and the call centers. Such channels need to be allocated a higher proportion of the fixed expenses. All these profitability measures need relevant accounts and customers to be available within the data warehouse. Moreover, profitability needs to be analyzed over time and sufficient historic data will be required.

Event Triggers

Another important tool that can be used for analytics is event-triggered campaigns. Campaigns can be based on specific customer actions (non-availability of credit limit, significant high cost services usage, etc.).

Other Analyses

There are several other analyses that can be carried out using BI tools, as follows.

Fraud Analysis

Transactions can be used for fraud detection. Such analysis uses historic data about customer, usage, payment record, and so on, that can be validated within the customer life-cycle framework and fraud detection triggers can be constructed.

CRM Components

A lot of the historic data within the data warehouse can be used to support the telecommunications company's CRM initiatives. A data warehouse provides a 360-degree view of a customer and enables a telecommunications company to study and reasonably predict customer behavior. A data warehouse integrates well with all campaign channels and provides a framework to generate integrated campaigns.

Predictive Models

A very important aspect of a data warehouse is its ability to provide integrated and historic information on customers, accounts, transactions, delivery channels, and the underlying data. All these can be used by a data-mining group to understand customer behavior over time, carry out trend analysis, and construct statistical predictive models. Models can range from attrition models to other predictive models that determine the probability of a customer's behavior.

STRATEGY AT WORK

A BI solution can jump-start the required impetus to a telecommunications company's strategic efforts. It is important for a telecommunications company to sequence these monitoring, organizing, and analysis efforts, in order to effectively utilize various resources. A great deal of time needs to be devoted to planning the various initiatives to maximize benefits and major coordination between various departments that are involved in these initiatives is required.

There are several critical areas and tools that any telecommunications company should undertake, although each telecommunications company might differ in the method of analysis and implementation.

Strategic Decision Support

This is the cornerstone of business intelligence. In this model, end users are provided with intuitive tools to distill information about corporate assets and their performance. Corporate assets include customers, products and services, network infrastructure, and employees. Typical performance measurements include profitability, availability, usage, sales, and lifetime value. Companies can now track key performance measurements, refine customer segments and scores, and optimize campaign strategies.

Some of the typical strategic decision support capabilities in the telecommunications industry include the following:

- Develop simple reporting capabilities that allow one to measure and trend key performance metrics; these metrics include the following:
    - Install and disconnect rates
    - Call-center average sales per hour

- Call-center average talk time
- Campaign performance
- Customer segment lifetime value
- Peak network volumes
- Uncollected receivables
- Customer satisfaction

- Develop complex reporting capabilities that allow one to uncover problems and discover new opportunities; typical areas for analysis include the following:
  - Market assessment
  - Channel planning
  - Competition assessment
  - Strategy and pricing
  - Customer penetration and profitability
  - Customer segmentation
  - Program definition
  - Recognition of patterns relative to customer behavior and needs
- Develop statistical models that predict customer needs and behaviors; for example, one can build models that predict a customer's likelihood to do the following:
  - Buy a new product
  - Generate high profitability
  - Respond to contacts through specific channels (e.g., direct mail, telemarketing, e-mail, etc.)
  - Not pay their bill

In addition, models can be built that predict network growth and fraud based on traffic patterns in the network.

Scoring and Segmentation

These provide the mechanisms for deploying score and segmentation rules developed through strategic decision support. Scoring provides processes that apply statistical models to each customer (or prospect). A score from 1 to 100 is then assigned to indicate how well the customer fits the model. For example, suppose that a model predicted who was likely to be a high-usage customer. This model would be applied to each customer and a resulting score would be assigned.

A score of 100 would indicate a near-perfect match to the model, as opposed to a score of 1, which would indicate that the customer did not fit the model at all. Segmentation provides a means for grouping similar customers. For example, one may segment the customer base between residential and commercial markets. In addition, one may decide to provide further granularity by defining segmentation within these subsegments.

Defining customer segments is the first key step toward defining a customer management strategy.

Campaign Assignment and Management

These start where strategic decision support and scoring and segmentation leave off. Now that we understand what products to deliver, to whom, and how, it is time to set up a campaign to orchestrate the contact activity. Generally, campaigns contain six key elements:

1. The list of customers to be contacted as part of the campaign
2. The channel to be used in reaching the customer
3. The product, program, and service to be offered
4. The incentive to be used in selling
5. The relationship relative to other campaigns
6. The priority relative to other campaigns

Once the campaign has been defined, it is executed via the contact management capability.

BI IN ACTION

BI can be a very effective means of analyzing, organizing, and monitoring the complex barrage of information generated in one's business and helping to generate a more effective business model for increasing revenue by keeping one's customer base happy and increasing profitability by cutting costs.

Customer Retention

BI tools can be applied to a variety of processes forming the telecom service provider's business. These business processes can be customer retention, cost cutting, or traffic management. For customer retention, strategic decision support BI tools would be used to track key performance metrics relative to customer install and disconnect activity and would assist tele-communications companies.

- This would provide early warning of increasing disconnect activity.
- If disconnect activity began to grow beyond acceptable limits, it would analyze why customers were disconnecting and extrapolate the impact on profitability.
- If the profitability impacts were not acceptable, it would formulate strategies for retention.

Once strategies were formulated, it would develop predictive models that would align retention strategies to the appropriate customers. Scoring and segmentation BI tools would assist telecommunications companies by applying predictive models from strategic decision support BI tools to the entire base of customers, assigning a score value.

The campaign assignment BI tool would assist telecommunications companies by applying scores from the scoring BI tool and other relevant data to assign customer lists to the appropriate retention campaigns.

Business management would initiate these campaigns and manage their execution. As feedback is returned from business management, input would be used by strategic decision support to refine retention strategies.

As may be deduced, a number of capabilities are needed to support a single business need (e.g., retention), and these capabilities are integrated through the business process. What may not be quite as evident is that these capabilities can be reused to support other business needs, such as customer care or fraud. Capabilities are essential to providing telecommunications companies with the ability to respond to the changing needs of their customers and the marketplace quickly and cost-effectively.

DISAPPOINTMENTS FROM THE PAST

Business intelligence solutions have had rather moderate success in terms of adding true business value to telecommunications companies. Notwithstanding the large investments in building BI solutions, telecommunications companies continue to face serious challenges in accessing data that is trustworthy, complete, and accurate, and data that makes business users self-sufficient.

Indeed, it is not uncommon to come across telecommunications companies where marketing and finance departments yield two different results on a critical input such as the success rate of campaigns. Also, if one adds up the time business users spend attempting to learn minute technical details in order to be self-sufficient and the costs of different departments acquiring their own technology pool to support their respective business needs, the drain on a telecommunications company's resources could be considerable.

On the other hand, though, technology departments tend to believe BI solutions have been extremely successful. Essentially, this is because technology teams assess success in terms of a very large database being implemented or a complex ETL (Extraction, Transformation, and Load) problem being addressed within the telecommunications company, as is frequently the case when a BI solution is implemented. Therefore, although BI solutions may have been technological success stories, their true value in terms of enhancing the trustworthiness of information or providing a holistic view has been limited.

BARRIERS TO BI

BI efforts are taken up in some cases at the enterprise level and, in other cases, at the department or function level. However, when building a BI solution, a common framework is usually adopted. The typical features of this framework are:

A common data repository with simplified structures to facilitate its consumption, periodic acquisition, and refining of data from many sources, and loading this data into the repository, and extraction of data from these sources with a set of reporting tools.

BI solutions built using this approach have grown exponentially in size and scope, forcing the creation of more manageable subsets of data to suit the business requirements of different departments. Even as these smaller subsets of data became de facto sources, seeds for their uncontrolled proliferation were sown and, today, telecommunications companies have thousands of these smaller subsets of data that business users depend on for their day-to-day analysis and reporting. With this as the context, let's examine some of the challenges telecommunications companies face today with respect to their business intelligence infrastructure.

I Do Not Get the Full Picture

Telecommunications companies find it extremely difficult to acquire a cross-functional or holistic view of data. For example, how can a risk manager combine the profitability view of customers with their risk view, and analyze dependencies between profits and risky behavior, a common business need for managing the risk of a portfolio?

Cause: Risk data and profitability data reside in silos. The basic definitions, data structures, and granularity of representation are different. It is a systems integration nightmare to combine the two views of data.

I Do Not Trust the Data

The same question often elicits different responses from two different departments of a telecommunications company. Which is the data that the senior management should trust and why?

Cause: Answers come from different silos although both silos may have obtained their data from the same underlying data repository. Different definitions: what is the process to ensure that the silos are leveraging the same definition or calculations? No traceability: how were the estimates and figures arrived at? What were the bases? Can these figures be traced back to the source data? What were the transformations along the way? Invariably, a telecommunications company would draw a blank to all these questions.

I Am Not Empowered

Can a user gather simple business intelligence on, say, the total number of new customers or the customer attrition figures as of a specified date? Very often, this may be an involved exercise. Again, to gather this information does a user have to be familiar with multiple query languages, tools, reporting interfaces, and databases? The answer from most business users is usually in the affirmative.

Cause: Users need to know the silo or data source to go to. Most users depend on "techies" to find them the answers. Most reporting tools require a basic understanding of syntax. Business users, again, depend on technical resources to delve into databases for relevant, often critical information. There is no common framework for leveraging definitions, and calculations reporting tools do not provide a common and consistent language for interaction.

I Cannot Close the Loop

To act, managers and analysts need current information proactively. For example, up-to-date information of high-end customers is critical to a customer relationship manager in determining customer satisfaction and probability of attrition. However, this kind of information seldom reaches line managers on time and proactively. The ability to "close the loop" by sending proactive information, and forcing the telecommunications company to act is a challenge that is mired in technological complexity.

Cause: Reporting solutions lack comprehensive alerting capabilities and are not flexible enough to allow users to define alerts themselves. Even if these facilities are available in some reporting solutions, business users require technical assistance to extract the information they require. Applications do not provide access to common business definitions and calculations that would allow business users to define seamless alerts and have the information delivered on multiple touch-points.

It Takes Too Long to Build a BI Solution

Despite considerable spending on a BI infrastructure, most telecommunications companies find that what they have does not match up to their unique business needs. Building a BI solution for a telecommunications company requires extensive understanding of the telecommunications domain and how business users perform various analyses.

Cause: The effort to build a BI solution frequently begins from scratch, with teams trying to understand business needs and building suitable models. This can consume significant resources and time. Teams building a BI solution lack domain expertise. Often, teams of modelers and architects spend too much time understanding the business. Most vendors offer tools and technologies for point solutions. There are few complete, connected, and consistent analytic solutions.

OVERCOMING BARRIERS

Building a BI solution is more than an implementation of a bunch of technology tools. It requires a framework driven by a set of guiding principles. Some of these core principles of implementation are explained here:

1. *Establish the need for trustworthiness:* A BI solution has to deliver data that is trustworthy. Data becomes trustworthy when it can be substantiated. Ensure that data that is delivered to business users can be substantiated with a seamless traceability all the way to its source, without any additional effort. When a business user looks at a number and wants to authenticate it, the task should be as simple as a click of a button. Users should not have to depend on technical resources to draw out simple reports. In short, the traceability and lineage of data should be comprehensive and seamless in a BI solution.

2. *Establish the need for a holistic view:* Build a BI solution to provide a holistic view. Even though the effort may begin with a single department or division or a specific subject area, the solution should be designed such that additional subject areas can be added seamlessly. For example, a telecommunications company can begin by building the BI solution for risk. However, the telecommunications company must ensure that the underlying risk structures are created such that profitability, CRM, and the investment sides of the business can be accommodated at a later date. If efforts are decentralized and managed by individual departments, it is important to ensure that they coordinate the underlying structural aspects in ways that can facilitate extensibility and integration. Unless such strategic guidance is available, efforts tend to be independent.

3. *Establish the need and sanctity of consistent business terms:* A BI solution without this fundamental driver is set to fail. Facilitate and create a glossary of terms that is consistently applied across your enterprise. It is important to involve business users in this exercise. Also appreciate the diversity between departments, and ensure that everybody's requirements are accurately described.

4. *Enforce consistency at all costs:* Defining or establishing a business language in itself does not solve the problem. A mechanism to enforce this common language must be established. The ability to enforce this feature across the BI solution is a key component of the solution. Telecommunications companies must realize that departments would prefer managing their own subsets of data and would like to limit their access to those subsets. Departmentalized subsets are inevitable, and are simple to manage and maintain. The challenge is to control their proliferation.

5. *Empower business users:* The ultimate test and one of the most uncompromising success factors of a BI solution is to empower business users and reduce their dependence on technology resources significantly. True empowerment comes when a business user interacts with the solutions using common and simple business language without worrying about where the data is stored and how the data is retrieved.

6. *Adopt connected and consistent solutions:* Guard against the tendency to go in for a point solution that supports your immediate business demands. When the risk department, for instance, chooses a specific ALM solution or an investment management solution without considering the need for holistic enterprise risk and CRM solutions, the scope to extend the point solution is limited. This is because point solutions are built for specific needs using varied technology paradigms and choices. Prefer, instead, an integrated, connected, future-proof solution: a solution that offers connected and consistent analytics that are derived from the same underlying business model, business definitions, and processing backbone.

7. *Reduce the implementation cycle:* Unless managed and controlled well, a BI solution implementation, particularly at the enterprise, division, or department level, can easily spin out of hand. The best way to avoid this is to adopt prebuilt solutions that are, nevertheless, extensible. A good rule of thumb is to opt for an analytical solution suite that can jumpstart your BI initiatives by 50 percent or so. This immediately reduces the implementation turnaround by about 60 percent.

8. *Close the loop and deliver proactively:* Another key feature of a successful BI solution is the capability to deliver business data proactively to users across multiple touch-points. Business users should be able to define the necessary criteria for alerts and the BI solution should be capable of sending information as alerts on a variety of touch-points. It is this that really determines the effectiveness of a BI solution: when information is offered in a manner that results in action.

Very few telecommunications companies have built BI solutions that have truly empowered them at all levels—strategic and tactical—to gain insight into their health and to act on this insight. Many an implementation story that is touted as a "model" centers on the technological rather the business aspects of BI. For a BI implementation to succeed in the business sense, it is important for top management to be committed and supportive, while providing the necessary direction and guidance. Also, it is important to adopt a strategic design and a common framework across the enterprise, implementing the solution in phases.

Furthermore, the management must keep in mind that point solutions that cater to specific functional needs may indeed be successful in the short term but, over the long haul, this may put the enterprise at considerable risk. Choosing prebuilt analytical solutions that are connected and consistent will improve the chances of success enormously.

CUSTOMER INTELLIGENCE

Out of 10,000 customers an organization has, 1,000 customers are providing 80 percent of revenue and many of these most profitable customers reside in a sales region handled by that sales manager recently hired. Think of how the organization's strategy might change if it knew some simple, basic facts about its customers.

Enterprises should see customer activity—interactions with customer service, accounts payable, sales and marketing, and more—in its entirety. In recent years, businesses have frequently ignored, or at least paid only cursory attention to, one of the most fundamental keys to success: their relationship with their customers.

The sometimes paradoxical relationship between customers and businesses arises from fundamental spheres of influence. Consumers had little influence on how businesses responded to their needs, and businesses could derive little reward by distinguishing themselves through customer relationships and superior service. In effect, businesses dictated the relationships with their customers, and customers often accepted that standard. The status quo reigned. In today's competitive environment, the nature of customer relationships has changed. Consumers have many choices to meet their needs, and aggressive advertising or access to the Internet increasingly broadens a consumer's horizons for competing products.

Companies are competing for the same customer, and successful businesses must provide a superior relationship with customers to stand out. In a way, the rise of CRM systems and methodologies that exploded in the late 1990s was merely a desire to return to "traditional" customer relationships. Successful corporations win and keep customers and prospects by establishing direct, sustainable, and manageable relationships.

Customer Data in the Telecommunications Industry

The telecommunications industry is a good example of how vital customer knowledge is to compete more effectively and nurture customer loyalty. Business managers and decision makers are constantly facing the need to answer fundamental business questions such as, "Who are our most profitable customers?" "What makes them profitable?" and "Which marketing campaigns should be developed to target this segment?" in order to monitor customer base evolution and behavior, define the most adequate marketing strategies, and assess the overall operational performance.

However, the intelligence required to answer this type of question is spread across several operational databases (billing, CRM, mediation, provisioning, ERP, etc.) that do not support the ability to easily query, report, and analyze business data. Moreover, it is extremely complex to cross data coming from different systems. For instance, correlating traffic and customer data would be very useful to define the most competitive pricing plans.

Customer Relationship Management (CRM)

CRM is the alignment of business strategy, organizational structure and culture, and customer information and technology so that all customer interactions can be conducted to the long-term satisfaction of the customer and to the benefit and profit of the organization. Implement a coordinated, customer-focused business strategy. An organization must have business strategies that promote CRM across functional boundaries. Goals that include phrases such as "customer-focused" or "customer satisfaction" are indicators that CRM is important. However, if there are no underlying strategies in place that force a customer view across business functions, the organization is not likely to move far from the traditional product focus.

Telecommunications companies having the desire for customer-focused initiatives must implement enterprise strategies for moving the company in that direction. It should create a CRM-friendly organizational structure (see Figure 7.4). The overall organizational structure must promote cross-functional cooperation. Independent product-oriented business units, multiple marketing and sales organizations, and distributed customer care centers can all inhibit an organization's ability to determine and carry out the next promotion or service activity for the customer. With the autonomy and control possessed by each business unit executive, the telecommunications companies may lack the organizational structure required to implement cross-department initiatives.

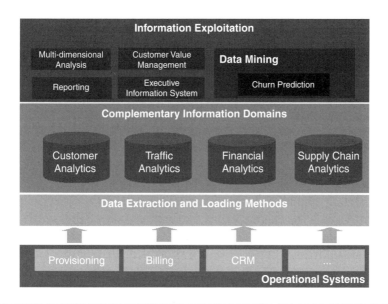

Figure 7.4 Customer information system

CRM-Savvy Organizational Culture

Culture is a critical but often overlooked factor that can have a strong influence on the success or failure of any CRM endeavor. There are three predominant aspects to consider: (1) the organization's ability and willingness to effect change to business and thought processes, (2) the degree to which the business units work together, reach compromise, and facilitate shared strategies, and (3) it is important that executives support CRM.

Integrated Customer Information Environment

Customer information is the cornerstone of a successful CRM program. This information must provide a common customer view and must be distributed across the organization to facilitate both operational and analytical uses. This always requires a technology architecture that integrates multiple applications, ranging from operational legacy applications to call center systems to the data warehouse and its associated data marts.

Although the technologists in the telecommunications industry understand the need for consolidated customer information, the business units of most companies across the industry until recently maintained their disparate systems. However, today's organizations are leveraging BI in conjunction with CRM solutions in numerous ways to derive hitherto unseen benefits and business opportunities. Using BI tools, the intelligence can be built into CRM applications as well.

Another point in favor of a BI–CRM combination is the fact that many organizations already have a CRM solution in place. It makes more sense for such enterprises to leverage existing CRM investments and deploy a full-fledged BI system. This will ensure that data from all the data sources (including the CRM) can be brought under one roof for a better all-round perspective of the business.

Using Client Data for Intelligence

Before you can establish meaningful relationships, however, companies must be able to answer—with precision and confidence—one seemingly easy question. Exactly who are my customers? In an increasingly competitive world, using the client database smartly, to gain a better understanding of the organization's number one asset, customers, can make or break the success of the organization. BI answers questions such as the following:

1. Who are my most/least profitable customers and products?
2. To whom should I address my marketing action/campaign?
3. What are the sales performances of this period with respect to my objectives and with respect to the same period last year?
4. Which are currently my best performing products in Germany?
5. Which customers are about to churn; who are fraudulent?
6. At what price should I sell my service/product in Geneva?

Most enterprises use databases to store information about their current customers, previous customers, business partners, and potential customers. The challenge lies in finding a way to harness the useful information contained within these high-volume databases in order to produce intelligent business solutions.

Analyzing the information that an organization stores in connection with all customer interactions can reveal a lot of remarkable facts about the buying behavior of customers, what motivates them, and what might make them stop buying from you. It also provides a scientific method to monitor an organization's own business performance.

Detailed analysis of customer data will also provide insight into their needs and wants. The exercise will analyze and segment customers' buying patterns and identify potential services that are in demand. Organizations can use this information to shorten response times to market changes, which then allows for better alignment of products and services with customers' needs.

An in-depth understanding of customers, provided through comprehensive data analysis, will also allow picking and targeting better prospects, achieve a higher response rate from marketing programs, and at the same time identify reasons for customer attrition and create or alter programs and services accordingly.

An important point to consider is whether the analysis is guided by predefined questions. Predefined points of analysis are aimed at understanding certain types of behaviors by analyzing relationships between various predecided influencing factors. For example, a predefined analysis of customer service V's sales would illustrate the effect of good and bad customer service on sales, and would answer questions such as how important customer service is to customers and how much it influences future sales. On the contrary, the objective of an open-ended analysis is to discover trends that are not anticipated by ordinary immersion in day-to-day business. Performing an open-ended analysis internally is often impaired by the expectations brought on by individuals working within the organization.

The techniques used to analyze data are complex. In order for an organization to be able to use the results of the data analysis, it is crucial that the results should not be clouded by the complexity of the calculations but are delivered in a straightforward manner. It is important for an organization to recognize that a good understanding of its customers is useful only to the extent to which this knowledge can be translated into real business practices. Business intelligence refers not only to the data analysis in itself, but also to how you relate the results from the data analysis to everyday business decisions and how you translate the recommended actions stemming from the analysis into live campaigns.

It is therefore important to ensure that the marketing department interacts with the data analysts constantly throughout the process. That way, when the data analysis is complete, the marketing personnel will already be in tune with the issues the organization is facing, and will be able to develop campaigns to capitalize on opportunities and strategies to mend weaknesses quickly and effectively.

Understanding how external market conditions affect your business will enable you to react quickly to future changes in the market. Finally, understanding customer behavior and the way customers use products and services will enable the organization to improve its service to its current client base as well as to target new business more effectively.

True customer intelligence is an elusive goal, but it is becoming a reality for companies tired of the incomplete and slippery information that tries to pass as corporate truth. For years, companies of all sizes in every industry have searched for ways to discover the truth about their customers. But more often than not, businesses face uncertainties about even simple truths, such as how many customers they really have and which ones are worth keeping.

In fact, the sad truth is that many large enterprises still do not know their customers very well. In most cases, businesses have amassed wads of customer data. As organizations increasingly standardized on different data collection methods—CRM, Enterprise Resource Planning (ERP), DWs, and the like—customer data often was replicated in different systems. And each business unit or division may have its own systems. This viral spread in applications led to a confused and untenable view of the customer.

To maintain, manage, and track these critically important relationships and the associated customer activity, corporations are investing valuable time and resources into managing customer data with Customer Data Integration (CDI) systems. As a result, every department has a different piece of the customer puzzle and the business has no way to fit them all together. Enter customer data integration, or CDI. CDI will change all this.

Customer Data Integration (CDI)

CDI is the process of consolidating and managing customer information from all available sources, including contact details, customer valuation data, and information gathered through interactions such as direct marketing. Properly conducted, CDI ensures that all relevant departments in the company have constant access to the most current and complete view of customer information available. As such, CDI is an essential element of CRM.

By applying rigor to the way companies locate, define, match, model, transform, and store their customer data, CDI solutions streamline heretofore labor-intensive, specialized, and manual data integration activities. CDI introduces the concept of a centralized customer data "hub," a single repository that can serve as the company's de facto customer system of record, thereby addressing a wide array of business needs.

The value of CDI is that it provides a "360-degree view" of the customer by bringing together data from disparate sources throughout the enterprise. A central component of CDI is an integrated master customer reference database with built-in data quality, correction, and correlation capabilities.

CDI is not only a boon to companies struggling to deploy CRM, it offers a single source for customer data that is leverageable across applications, systems, and knowledge workers. I liken it to the change machine at a grocery store. It takes the loose data jangling around the pockets of your organization and, for a relatively small fee, converts it to hard currency.

With CDI, for example, a customer can walk into a bank and the teller can instantly check that customer's profile across other departments such as insurance, financial planning, and credit cards. The teller can then identify whether that customer would be an ideal candidate for additional product offers.

Technology can solve a lot of integration problems, but there is still no silver bullet for enterprisewide data governance. That is why companies are overhauling data management on an "entity-by-entity" basis, rather than diving into huge Master Data Management (MDM) initiatives; companies are first addressing customer data integration and product information management separately.

But, even for companies simply taking on CDI, there's a lot to consider: varied technology approaches, high implementation costs, and a market that's changing quickly. The toughest part about master data management is the governance: getting everyone to agree who owns the data, how that data is manipulated, and who's in charge of cleaning up the information.

Alternate Approach to CDI

One approach that is gaining momentum is to start CDI efforts with a registry-style system while working through the complexities of implementing an operational-style hub. Registry-style hubs store cross-referenced links to the various data sources in an enterprise, whereas operational-style data hubs store actual master customer information and reference data in a central location. Registry hubs generally have faster deployment times, and hence more immediate ROI, but there is a downside: applications lose a lot of the richness of a (operational) data hub.

So, vendors are offering more flexibility and "richness" by developing harmonized or hybrid CDI systems, incorporating features of both registry and operational data hub styles. Registry vendors will start to look more like operational data hubs, and operational data hub vendors will work to speed up ROI for their products. Ultimately, the real winners will be vendors that can navigate the underlying data model challenge, inherent to CDI success.

The data model issue will remain a hot CDI topic for vendors and companies. It will play out with early adopters choosing a more flexible approach, followers likely choosing market-tested models, and vendors solidifying their data model messages.

Benefits of CDI

CDI is a combination of technologies and processes that manage the integration held within customer information systems so that interactions can be managed for the mutual benefit of both the customer and the business. The ultimate success of a relationship between a business and a customer is determined by the quality of the interaction. Successful CDI results in:

- Significantly enhanced customer service by understanding what the customer needs
- Increased customer satisfaction by providing timely informed options
- Higher customer retention as consumers view the company not as a vendor but more as a trusted provider of goods or services

- Lower cost of acquiring customers by using aggregated data sources to refine sales and marketing messages
- Better understanding of customers, leading to better decisions in product offerings, enhancements, and packaging
- Reduction in duplicate critical customer information, leading to improved marketing campaign results and sound forecasting practices
- Improved business intelligence reporting by providing more accurate data to reporting applications, leading to more timely, accurate reports to decision makers

CDI also promises to help organizations better comply with federal regulations such as the Can Spam Act and "Do Not Call." Armed with a complete and accurate customer record, marketers and call center agents are far less likely to make costly compliance mistakes.

Customer data integration systems are complex puzzles with many interlocking pieces, where each individual piece serves a purpose. However complex and detailed the individual pieces are, the CDI puzzle is not finished until all the pieces are integrated and the picture is complete. But reconciling data from heterogeneous and often geographically disparate systems remains one of the biggest challenges.

CDI Challenges

Key customer data integration challenges that prohibit a company from reaching its operational performance are:

- Customer data is created and locked in multiple silos including applications, data warehouses, data marts, and external sources. Applications such as customer relationship management and multi-channel touch-points such as Web sites, call centers, and e-mail have created an environment where information of a single customer is strewn across scores of database silos in different lines of business or product divisions. Worse, the same customer information is duplicated by each application, recreated differently for each business process, or stored repeatedly in various data warehouses.
- Customer data is owned by different parts of the business, across organizational boundaries. Data unreliability stems from inaccurate and inconsistent customer reference data duplicated across the various IT systems within an enterprise. Different lines of business and business units have their own systems and data sources that contain customer information that is relevant to their business, but often contains reference data that is in conflict with other customer reference data in the organization.

- Many of these data sources have proprietary fixed data models and standardizing on a single application vendor model is not feasible. It is extremely difficult to change or extend the data model that comes with these applications, which makes it challenging to consolidate data from other applications and data sources that are outside the scope of the application vendor. Most enterprises have customer data spread across 20 to 30 data sources, and applications from a single vendor typically control at the most 10 to 20 percent of these sources. The rest of the data sources, including external data sources, have to be mapped and transformed to feed data into the vendor's data model. Therefore, standardizing on the application vendor data model means more, not less, work. Second, because the application vendor data models were developed originally to support proprietary applications, these models are too complex and contain too many extraneous attributes that are not required for a CDI solution. This not only duplicates a lot of unnecessary data in the customer hub and impairs the performance of the system, but also makes data mappings, data imports, and addition of new data sources unnecessarily complex.
- Heterogeneous architectures and tools (i.e., "IT stack") have different data integration technologies. As data silos break down, increased connectivity and integration efforts reveal new data incompatibilities in data modeling, data semantics, data quality, and data management.
- Legacy data hubs, such as the Customer Information File (CIF) are not easy to extend, nor can they be abandoned. Organizations with legacy data hubs have long understood the value of clean, accurate customer information, but have been frustrated by the limitations of older generation CIF systems, which required them to adhere to a stricter product focus rather than customer focus and lacked flexibility to blend data from third-party sources. Yet, these legacy data hubs still contain valuable historical customer information that must be accessible within the new customer data integration platform/infrastructure.
- Most data tools work only in batch or real-time, not in both (e.g., EAI and EII tools are designed to work in real-time, whereas ETL is designed to work in batch). Each of these tools was originated for narrow purposes and is ill-suited for CDI: ETL was developed to move large volumes of data in batch mode, whereas EII was developed to run distributed queries across disparate sources in real-time. As a result, each of these options effectively supports only a single data modality; batch or real-time, not both. Because customer data is inextricably tied to both operational and strategic business processes of a company, such as the order-to-cash process or profit-

ability segmentation analyses, it needs to be delivered in time for each business process and real business.

■ Significant data quality issues and conflicting semantics (metadata) exist within and across data sources. Although there are transactional systems of record (e.g., billing), there is no single reliable "system of record" for representing customers' profiles within the enterprise. Customer profile (reference) data is distributed across the various IT systems within an enterprise, with large amounts of duplication and a high degree of inconsistency and inaccuracy. In addition, how one operational application describes a customer may be very different from another operational application, leading to semantic inconsistencies and intensifying customer data reliability issues.

LOOK BEYOND THE ORGANIZATION

When most people think about populating their customer information file (whether it is a data warehouse or operational data store), they look internally at their customer databases. These sources—typically management systems, call center systems, and sales systems—contain information the organization has about each customer and each customer's touchpoints with the company.

Although these sources of information are absolutely critical, they do have a major deficiency: they only contain information that the company has collected about its customers. That doesn't mean it isn't useful information. With this information, for example, the company can analyze its sales and establish some trends. It could identify pairs of products that sell well together, and then consider customers who only purchased one of these products and target those customers as prospects for a compatible product.

The company could be much more effective in its marketing efforts if it could augment or enhance its customer data. There is a variety of external data sources that can help here, and we describe some of them in this book.

Personal Demographic Data

Personal demographic data is a set of characteristics about each customer or prospect of interest. These characteristics include age, household (and personal) income, marital status, number of children, credit card debt, home ownership, and net worth. Here are some critical business questions that could be answered using this data.

■ What are the buying patterns of people in a specific income bracket?
■ How do sales patterns change as people migrate from one age group to another?

- Which products sell better to homeowners and which sell better to renters?
- What are the characteristics of customers who buy certain products?
- Who are the other customers who share these characteristics and do not buy these products?

Each of these questions yields a group of prospects for which a directed sales campaign can be targeted. Companies that use such information in developing their direct mail (and other) campaigns are more likely to have a higher success rate than those that do not. The higher success rate occurs because the people who receive the mailing are those who are most likely to buy the promoted product. In addition, by screening out people who don't fit the profile, the company is not wasting time and money and is not bothering customers with offers in which they are not interested.

Geographic Demographic Data

Geographic demographic data provides similar information, but instead of providing it on individuals and households, the information is provided on geographic areas such as census tracts, postal codes, and municipalities. By knowing where the existing customers live, the company can then extrapolate this data to obtain (likely) characteristics about its customers. Some caution is needed in this case, because individuals within the geographic area may be exceptions. For example, although the geographic demographics for an area indicate that it is primarily populated by young professional adults, some of the people living in that area may not fit that profile.

Geographic demographic data is also useful for supporting geographic analysis. With internal data, the company can identify distances. For example, it can discern that its customers travel to the retail outlets. Armed with this information, the companies can also perform an analysis to see if the travel distances differ based on the characteristics of the geographic areas. Retail companies analyzing sites for future stores cannot only estimate the cannibalism (existing customers migrating to their new stores), they can also estimate the compatibility of their offerings with people living in certain geographic areas.

Attitudinal Data

Customers have opinions about the company's products and services. Often, these opinions go unnoticed except when the company receives a complimentary letter or a complaint. To better understand its customers, a company could conduct (or engage an external firm to conduct) a customer survey. A properly structured and administered customer survey can provide a wealth of information that can be used to adjust the product, its delivery, the associated services, the fees, and so on.

Deployment Considerations

Combining internal and external data is not always a straightforward process. At the customer level, the internal customer record must have attributes that are available for the customer in the external database. Without common information, the customers in the internal database cannot be reliably matched to people in an external data file. Examples of such attributes that can be used for matching include social security number, phone number, and address.

Similarly, customer opinions are of limited value unless the opinions can be linked to customer clusters. Comments by new customers provide information that is helpful for customer retention, and comments by unprofitable customers provide information that could possibly be used to turn them into profitable customers. Hence, it is important to obtain matching data in the attitudinal survey information as well. Companies should customize the survey by target group or include questions that could be used to link the responses to customer segments in the survey.

Education is another key deployment consideration. Through the external databases, the company will be learning a lot about its customers. The people using this information need to be educated in what they can use, what they can reveal, and to whom they can reveal it. For example, if we learn a customer's birth date through an external database, it may be tempting to acknowledge the customer's birthday. If the company feels this is advantageous (e.g., the person may be eligible for some benefits), then this may be appropriate. There are circumstances under which a customer may be offended by the use of the birth date and misuse of the information could cost a valuable customer relationship.

External data sources, in addition to the internal databases, can significantly enhance companies' marketing capabilities. Armed with more complete data about customers and prospects, companies can perform more comprehensive customer analytics and can be much more effective in their marketing efforts.

ROLE OF BUSINESS INTELLIGENCE

Today's CDI systems have evolved into highly sophisticated applications incorporating leading-edge research and development advances in fields such as information theory, natural language processing, artificial intelligence, and others. One major advance has been the recognition of users' needs to be able to fine-tune the matching and householding behavior to create a single customer view that more directly fits with the business needs.

CDI vendors no longer assume that they can dictate to businesses what the "correct" single customer view is. As businesses have become increasingly sophisticated with business intelligence, CRM, and one-to-one systems, they have demanded control of their customer definition. This is typically affected via business rules that control how the single customer view is resolved by the CDI system.

The BI and middleware vendors will collaborate with CDI vendors to provide tighter integration. In addition, given the importance of trustworthy data to both BI and CDI projects, trusted data sources will play a bigger role in the market. With most organizations investing big money in such initiatives and also opening up the system to Web self-service, salespeople entering data, field service entering data, and call centers entering data, the data can get corrupted pretty quickly. Rather than argue over which department's data is right, companies will look to trusted data sources to append or be the source of record for customer information. These data vendors are offering more services to the CDI market today and, in the future, they may be more aggressive in the space.

While companies grapple with CDI, many are also implementing business intelligence tools, and these technologies will meet in the middleware.

COMPETITIVE ANALYSIS

Competitive Intelligence (CI) is a specialized branch of business intelligence and has become an important initiative in the present competitive scenario. Next to knowing all about your own business, the best thing to know about is the other fellow's business.

Competitive intelligence is defined as a systematic and ethical program for gathering, analyzing, and managing external information that can affect your company's plans, decisions, and operations. In other words, CI is the process of ensuring competitiveness in the marketplace through a greater understanding of competitors and the overall competitive environment. CI is not as difficult as it sounds. Much of what is obtained comes from sources available to everyone, including government sources, online databases, interviews or surveys, special interest groups (such as academics, trade associations, and consumer groups), private sector sources (such as competitors, suppliers, distributors, customers), or media (journals, wire services, newspapers, and financial reports).

The challenge with CI is not lack of information; it is the ability to differentiate useful CI from chatter or even disinformation. Of course, once you start practicing competitive intelligence, the next stage is to introduce countermeasures to make the CI task about you more difficult for other firms. The game of measure, countermeasure, and counter-countermeasure, and so on to counter to the nth measure is played in industry just as it is in politics and in international competition.

8

BUSINESS INTELLIGENCE: IMPACT

The bottom line is that the company with the most streamlined operations and the most reliable, useful customer service will become a market leader.

Over the past couple of years, most of the Business Intelligence (BI) vendors have relabeled their products as performance management solutions. However, their definitions of performance management have varied, depending on what products they happen to include in their suites with marketing departments emphasizing the demonstration of BI impact on improving profitability by reducing cost, enhancing operational efficiency, better risk management, and customer retention.

Although with all the hard work the proponents of BI are putting in, still it is very difficult to measure the impact of BI on business performance. Even without the impact of BI, just measuring the business performance itself is a tough job. It goes without saying that BI has been consistently delivering benefits in this vital area; however, many studies have found that "soft" benefits, such as improved reporting and better decision making, are much more likely to be achieved than "hard" benefits such as headcount reduction.

Two forces are driving the need for business intelligence, the massive increase in transactional data brought about by the increase in information systems and the requirement to understand and to analyze the business: financial results, sales results, and Key Performance Indicators (KPIs). Understanding a business meant analyzing the results and fine-tuning operations, which becomes more difficult as more and more data is generated. The increased pace of business means that companies must act quickly in order to successfully capitalize on new opportunities, which puts the ability to understand and analyze the data at a premium. Clearly, business intelligence is necessary to evaluate and analyze all the data that is generated, and to help find the key indices and trends that help people make better decisions.

In the past, business intelligence was hindered by a set of business and IT constraints that prevented a full transformational effect on business. Business Performance Management (BPM) is a convergence of technology that removes these hindrances through integration and enhanced technology to help streamline business transformation.

Successful corporate strategy has always relied on getting a straight answer to this question: "How's my organization performing today?" But how many organizations actually have the systems and processes needed to get a straight answer? In recent years performance measurement and management have received considerable attention from academics, practitioners, consultants, and policy makers. Many organizations have invested substantial amounts of time and effort in designing and implementing new performance measurement systems. Not many organizations have the ability to manage scenarios, accurately forecast, or have much, if any, forward visibility into future performance.

But things are changing rapidly, thanks in large part to BPM. Most of the organizations using performance measurement systems have a common view that the performance measurement system facilitated:

- Improved customer service
- Increased efficiency
- Broader focus on performance
- Instant feedback
- Greater staff engagement and motivation
- Improved communication
- Focus of attention on improving key issues

TELECOM CHALLENGE

In today's extremely challenging business environment, telecommunications companies are under intense pressure to reduce or eliminate the major threats to these slim margins: revenue leakage, churn, inefficient network usage, and least-cost routing plans.

Telecommunications companies rely on analyses of their terabytes of calldata record (CDR) data to help them make business-critical decisions that will positively affect their bottom line. High-end data warehouses and powerful business intelligence solutions are essential tools to help carriers meet profit goals.

Many critical telecommunications functions rely on fast complex analysis of CDR data. Key initiatives include analyzing behavioral data using CRM programs to optimally target services and reduce churn, ensuring complete and accurate billing and modeling call behavior with revenue assurance programs, and optimizing network operations using operations management programs. These initiatives all benefit from improved access to CDR-level data, access to large quantities of historical information for trend analysis and from the ability to quickly run complex BI queries.

These significant performance limitations force carriers to make a choice. They must either summarize or filter the data for analysis, or create a massive, complex, and often custom CDR warehouse to analyze call detail information. Both of these options pose serious limitations and challenges, resulting in incomplete information for decision making, or costly and time-consuming system development and maintenance.

There is tremendous value latent in call detail information for CRM (Customer Relationship Management), revenue assurance, fraud detection, and network usage analysis. Performing real-time analysis on voluminous call detail data with complex queries requires much more performance than legacy general-purpose systems can provide. Effective BI programs open the door for increased profitability, while eliminating the barriers to accessing and analyzing dynamic detailed information, and offering carriers performance, value, and simplicity in a data warehouse system. For the first time, carriers can leverage their terabytes of CDR data for real-time, better informed, and more strategic business decisions.

The Race for Performance

Every day more and more people are moving toward an alternative retail channel, the Internet, to get information or buy things. The impact of this phenomenon on business is clear, that is, shrinking profit margins as smart consumers compare shops with the click of a button and buy at the best value provider. The increased competition from new or unexpected areas and surging customer and partner expectations is putting unprecedented pressure on organizations to make processes more efficient and cost-effective.

The Internet's global reach and the impact of online business have also made people think that getting information should be easy. This is true for everyone from your employees who need information at their fingertips to do their jobs effectively, to customers who will look elsewhere if they don't easily find the information they need. The bottom line is that the company with the most streamlined operations and the most reliable, useful customer service will become a market leader.

Performance Measurement

Contemporary operating environments challenge organizations. Continuous improvement is important, thus for an organization to be successful, it is essential that an organization manage its competencies. In order to lead an organization, management needs information on the organization's current stage and the direction in which it is heading. The change has caused new managerial challenges and has resulted in the emergence of new management concepts and tools, such as performance measurement.

These concepts are expected to be better suited to the development of an organization's performance in a modern business environment. The new management tools have been developed to solve different problems from different managerial points of view. Simply stated, performance measurement deals with the implementation of an organization's strategy.

The main rationale for measuring an organization's performance is to be able to manage it. "If you don't measure it, you can't improve it." This statement cannot be truer than when applied to these initiatives, intended to revolutionize business processes. Performance measurement can be used as a tool for implementing an organization's strategy. Performance measurement can be used:

- To translate an organization's strategy into concrete objectives or goals
- To communicate the objectives to employees
- To guide and focus employees' efforts accordingly as these objectives are achieved
- To control whether the strategic objectives are reached
- To visualize how individual employees' efforts contribute to the overall business objectives

Performance measurement is usually carried out using a performance measurement system, which consists of several individual measures. There are many frameworks for constructing such a system. The most commonly used model is the Balanced ScoreCard (BSC). Others include, for example, the performance prism and the performance pyramid. The measures for the performance measurement system chosen are based on an organization's vision and strategy. Measures are chosen to measure success factors from different points of view, such as that of the customer, employees, business processes, and financial success, as well as from the point of view of past, current, and future performance. This way, different aspects of an organization's performance can be measured and managed.

There are four main phases related to the performance measurement process. First, the measures are chosen as described above. Then, the measurement system is implemented into the organization. This includes, for example, determining how the data for the measures is collected, how the measurement results are reported, and how the measures are used. The third phase is simply to use the measurement system. The final phase, the updating of the measurement system, closes the loop.

As crucial to the success of process improvement efforts as is the right selection of measures is the setting of improvement goals. Goals allow companies and individuals to measure success and progress toward it. When setting goals to accomplish revolutionary process improvement, one must make them aggressive.

There are several ways to determine what the right aggressive goal should be. One way is to set it at a level equal to the best process performance achieved in emergency or fire-drill scenarios. The reason for this is that in these situations, obstacles tend to be knocked down, and everyone pulls toward a common objective. Unfortunately, the obstacles go back up as soon as the emergency is over. Other ways to set goals include determining what it takes to meet customer expectations or what is needed to best the competition. Whatever method is used, remember that the objective is to cause a revolution in how business is conducted.

Selecting the right measure is directly linked to the criteria used to select and characterize the critical processes. That criterion is the need to improve the results delivered by the processes. Results are the first category of measures. They are the high-level metrics (EPS, revenue, return on investment, customer satisfaction, etc.) that are essential to the business. However, because they are at a high level, focusing improvement efforts on them is equivalent to trying to swallow too much. This is why a second category of measures, drivers, is the correct focus of improvement efforts.

Drivers are called such because they are the measures that, when improved, drive improvement in results. Drivers are selected at a level in the process that is actionable by the team tasked with improvement. They are designed to reflect changes made in the process in near-real-time. It is important to note that one driver measure is not sufficient to characterize process improvement, as one measure can be too easily manipulated.

For example, project cycle time can be shown to improve by reducing feature content of the product. However, features are essential to customer satisfaction. Therefore, both cycle time and feature content are drivers to be measured in this project example. In many cases, multiple drivers are required to have a truly comprehensive measurement program. Examples of drivers are time, yield, requirement stability, and cost.

Business as a Process

Having this information lets enterprises analyze their business on the fly, quickly detecting shifts in the market, and identifying new patterns in response to a promotion. The goal is to drive closer and closer to a real-time look at customer and sales activity. To survive, organizations must find new ways to leverage the Internet and make it easier for people to do business with them. This means creating innovative Internet applications that use legacy data in new profitable ways.

E-business has aligned the goals of business and technology more closely than ever before. To achieve synergy among all the systems that are operational in an organization, they must be able to directly and quickly integrate new Web applications into the existing corporate data and application mix, so that they can work together seamlessly as one. This means to simplify the business processes that make information access and distribution easy for people both inside and outside the organization.

Processes are the engines that produce the products and services of a company. A process simply involves the steps that must be accomplished to complete a task. For example, the steps in a process to develop a software product might be the following: identify customer requirements, design and code, test, install, and confirm customer satisfaction.

Obstacles exist between the process steps; overcoming these obstacles takes time and effort in the daily functioning of this process, effort that can be better spent elsewhere. When the obstacles are removed, the process flows smoothly. Transition from step to step is seamless and effortless. Obstacles do not exist, and the time to deliver the product to market is reduced.

To see a business as a process is to recognize that individual process steps are linked and dependent on each other for success. Dramatic improvement will only occur when the individual process steps and the links between them are the focus of improvement efforts. Often, the largest obstacles exist between process steps, at the links, requiring a holistic process view and cross-functional cooperation to effect improvement.

PERFORMANCE MANAGEMENT

The mandate to operate at an optimal performance level and meet organizational expectations is transforming the way organizations do business. This evolution is driving higher standards of competence in day-to-day operations and adding new pressure to increase stakeholder value, which can be achieved by applying performance management principles to technology investments.

Enterprises are searching for ways to bring IT investments in line with their most critical business priorities. This quest has created a shift from transaction-centric to information-centric approaches offering collaboration and information that can drive actionable decisions throughout organizations. Capturing knowledge and information across the organization can build long-term value for all stakeholders.

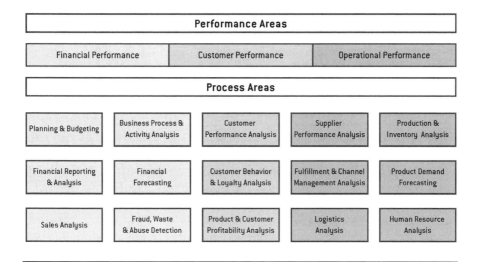

Figure 8.1 Performance and process management areas

In today's rapid-paced business climate, the agility with which an organization manages performance can determine market position and organization profitability (see Figure 8.1). Now, as never before, businesses have the ability to implement performance management to transform entire organizations with a focus to achieve goals. And now, as never before, the pressure is on businesses to manage their performance.

Performance management provides a critical foundation for organizations to manage their businesses and empower individuals to make the right decisions to maximize profitability. However, the present business environment also leads to some technical constraints and challenges for business performance management. These challenges, central to success of the performance management initiative, revolve around managing and analyzing high volumes of data in real-time while making available personalized relevant information to everybody.

Aim of Performance Management

The core principles and goals of performance management are the following:

Efficiency: The ability to optimize the operations and actions of the organization, individuals, and business processes to ensure that they result in defined goals and desired outcomes

Quality: The ability to continuously improve the quality of relationships, processes, and products or services to fully leverage practices or methodologies that maximize the value of resources and assets

Value: The ability to create and manage assets to increase business throughput and deliver long-term ROI for maximizing stakeholder return

BI THE CORE OF BPM

This focus on performance management and associated challenges is the impetus behind a new class of software capabilities now evolving from the roots of business intelligence and decision support that help organizations manage performance in a methodical and coordinated manner.

Business performance management links business intelligence to business strategies and processes via business metrics. This adds a top-down approach to the traditional bottom-up approach of business intelligence, and it adds new processes to the traditional business intelligence processes. These new processes are in particular about publishing and distribution of performance metrics and performance indicators to all people who have responsibility to run and to manage business processes.

The impact is that more and more information for monitoring, controlling, managing, and optimizing business processes is to be processed, because you can only manage what you can measure. Volumes of data warehouses and operational data stores are increasing as never before. And given the cycle speed of certain business processes, given the real-time opportunities of innovative channels such as the Web, mobile/wireless, and voice, business performance management must become real-time in many situations such as real-time trading and brokerage on electronic markets, real-time interaction management with customers, and real-time monitoring and controlling of trading, production, and transportation processes.

Performance management is ubiquitous: because everybody in an enterprise has responsibilities for executing or managing certain activities or processes, everybody needs metrics to measure and manage the performance of his activities and processes. The principle is "information democracy;" deliver the right information to the right information consumer to the right location at the right moment: an information supply chain paradigm is to be implemented. As enterprises have started to share key information, knowledge, and processes with its partners, suppliers, and customers, the information supply chain extends beyond the borders of an enterprise. Information democracy now includes all constituents of an enterprise's value net.

Process-orientation (collaboration, customer relationship management, supply chain management, business integration, etc.) drives Business Performance Management (BPM) for efficiently monitoring, controlling, and optimizing business processes via a closed-loop system approach (see Figure 8.2). Business performance management is the next generation of business

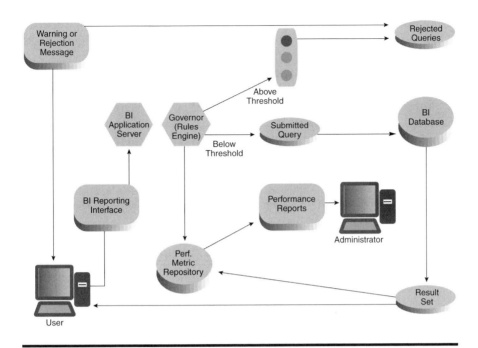

Figure 8.2   Generic BPM architecture

intelligence. It links business intelligence to business processes and business process management, and it uses the information democracy paradigm: availability and accessibility of the right information to the right information consumer at the right location in due time. From a technology point of view, there are still two open issues about business performance management:

1. Personalize information deployment to the information consumer via publishing, delivery, subscriptions, alerts, and notifications across all channels (Web, mobile, traditional).
2. Enable real-time access to distributed operational and analytical data without moving data into operational data stores.

Secured, personalized mass distribution of information according to the information supply chain paradigm is still a challenge. Local postprocessing of personalized information is a must, but nearly all vendors still include "real" data in information objects. The problem is limitation on size of information objects, insufficient speed of queries when objects get big, and quality problems in data consistency (risk of data manipulation by information consumer).

Enterprises are now taking steps into initiatives such as e-commerce, supply chain automation, and customer relationship management. Often this requires building custom solutions or buying new ones, such as packaged applications. And that results in a technology infrastructure that's piecemeal, disconnected, and difficult to manage.

These stovepipe systems sequester information from one department to another, making it difficult to achieve a panoramic view of the overall workings of the enterprise. Organizations end up with a constantly shifting, disparate environment. Somewhere in there is usable information to help customers make decisions, managers meet quotas, and engineers advance the capabilities of the company.

Today people constantly crave real-time knowledge about their business. Analysts who were formerly satisfied with weekly reports get impatient if they can't get 24-hour turnaround. They look at morning data in the afternoon so they can change their strategy slightly for the next morning. This holds true for everyone in the organization. The CEO wants a global real-time view of the entire organization, on demand. Sales managers are concerned with making quotas and growing the business into new regions. Engineers want to access current design specs and supplier information. Support people need to know how products are being used. Shipping clerks need to know with which couriers to do business.

To implement widespread knowledge of the business, organizations must make it simple for people to access the information they need to help them make better decisions. User preferences are also a factor. People want to see information in a way that makes the most sense to them, whether it's a spreadsheet, a bar chart, e-mail, or a Web browser. By simplifying the way people get their reports, you open up decision making to middle-level managers and staff workers throughout an organization. This allows companies to react more quickly to market changes and business problems instead of having to wait until the situation is filtered to upper-level management and delegated back to the front lines.

No matter how the information is delivered, it must be pulled together and integrated from different places—departments, business units, merger partners, and external public sources—and presented as consistent, accurate, and reliable information for anyone who needs it.

Business intelligence technology allows organizations to transform data stored in core business systems into meaningful information, which in turn enables organizations:

- To know their customers and let customers know them
- To streamline business processes by aligning technology with business goals
- To know how their business runs as a whole by getting a global unified view of the organization

UNDERSTANDING BUSINESS PERFORMANCE MANAGEMENT

Business performance management comprises business intelligence, Business Process Integration (BPI), business systems management, and collaboration (human interaction). These capabilities provide tight integration of operational and analytical environments, business and IT environments, and strategy with daily operations. Business performance management combines business processes, information, and IT resources, aligning the organization's core assets, people, information, technology, and processes, to create a single integrated view, with real-time intelligence, of both its business measurements and IT system performance. This integration of resources allows the organization to obtain business information faster, respond more quickly to market trends and competitive threats, and improve operational efficiencies and business results, all attributes of an on-demand enterprise.

Performance management is a process that connects metrics, methodology, and financial plans using technology, resulting in the right information getting to the right people, at the right time, in a way that allows for changes to be made. More important, performance and process improvement is an evolution, not a discrete action or one-time action. It is constantly evolving.

In that sense, BPM is a new solution to a very old problem. Indeed, despite the best efforts of technology and management gurus, visibility and connectivity into day-to-day operations have often been inconsistent or absent at even the world's best organizations. The boardroom remains disconnected from the front line, and vice versa. Strategy gets created in a top-down, ad hoc fashion that leaves average employees wondering.

BPM connects business users with the information they need to improve performance. That could be anything, from demand plan scenarios to regular updates of response time in the call center. All users need a view into the performance metrics, goals, and details that can be viewed in dashboards, scorecards, financial plans, and reports. Of course this requires a companywide effort to collect and prioritize metrics across business lines and functional areas. It also requires a common business infrastructure, to manage, secure, and control data from any and all technology infrastructure and source systems.

Business performance management is defined by a set of domains that creates a comprehensive infrastructure for monitoring, managing, and executing business performance information. These domains are:

- *Business rules domain:* Enables the exploitation of business rules for dynamic business process control and adaptive performance management
- *Workplace domain:* Delivers a collaborative workplace for users with specific role-based perspectives on performance management

- *Business systems domain:* Bridges the gap between business and IT for systems management, enabling a business viewpoint on IT operations and an IT viewpoint on business events
- *Process domain:* Provides tools and runtime engines for modeling, integration, management, and continuous improvement of business processes
- *Common event infrastructure (CEI):* Provides integrated enterprise delivery of business and IT events
- *Information domain:* Specifies the technical interfaces that partners can exploit to analyze and report real-time business events and performance information

Business performance management is a set of processes that help organizations optimize business performance. BPM is seen as the next generation of business intelligence. BPM is focused on business processes such as planning and forecasting. It helps businesses discover efficient use of their business units, and financial, human, and material resources.

Over the past decade, business performance management frameworks such as the balanced scorecard have been adopted by a growing number of major organizations. With user organizations showing high satisfaction, BPM frameworks such as the balanced scorecard will be a staple within large organizations for the foreseeable future. The business requirement for BPM frameworks manifests itself at all levels and across all functions of an organization. Many executives now recognize the need to manage business performance strategically, to understand the key linkages between strategy and the business processes undertaken to execute strategy, and to measure the performance of those business processes. This need for strategic performance management tools has been the impetus for the BI and BPM tools. At the same time, functional managers and middle managers need performance management tools appropriate to the business processes they manage, which, it is hoped, are aligned with the business strategy.

This need has given rise to different strategies for using balanced scorecards. These range from the traditional top-down approach with supporting scorecards "cascaded" down from corporate, to bottom-up approaches where business units devise scorecards that are meaningful for their purposes and there is no aggregated view of the whole company due to local differences in measures.

BPM involves consolidation of data from various sources, querying, and analysis of the data, and putting the results into practice. BPM enhances processes by creating better feedback loops. Continuous and real-time reviews help to identify and eliminate problems before they grow. BPM's forecasting abilities help the company take corrective action in time to meet earnings projections. Forecasting is characterized by a high degree of predictability which is put to good use to answer what-if scenarios. BPM is useful in risk analysis and predicting outcomes of merger and acquisition scenarios and coming up with a plan to overcome potential problems. BPM provides key performance indicators that help companies monitor efficiency of projects and employees against operational targets.

Metrics/Key Performance Indicators

BPM often uses key performance indicators to assess the present state of business and to prescribe a course of action. More and more organizations have started to make data available more promptly. In the past, data only became available after a month or two, which did not help to suggest to managers to adjust activities in time to hit Wall Street targets. Recently, banks have tried making data available at shorter intervals and have reduced delays, for example, for businesses that have higher operational/credit risk loading (e.g., credit cards and "wealth management"). A large multinational bank makes KPI-related data available weekly, and sometimes offers a daily analysis of the numbers. This means data usually becomes available within 24 hours, necessitating automation and the use of IT systems.

Most of the time, BPM simply means the use of several financial/non-financial metrics/key performance indicators to assess the present state of business and to prescribe a course of action.

Some of the areas where top management analysis could gain knowledge from BPM are:

- New customers acquired
- Status of existing customers
- Attrition of customers (including breakup by reason of attrition)
- Delinquency analysis of customers behind on payments
- Profitability of customers by demographic

This is more an inclusive list than an exclusive one. The above more or less describe what a bank would do, but could also refer to a telephone company or similar service sector company. What is important is that BPM integrates the company's processes with CRM or ERP. Companies become able to gauge customer satisfaction, control customer trends, and influence shareholder value.

BPM Tools

People working in business intelligence have developed tools that ease the work, especially when the intelligence task involves gathering and analyzing large amounts of unstructured data. Tool categories commonly used for business performance management include:

- OLAP—Online Analytical Processing, sometimes simply called "analytics" (based on dimensional analysis and the so-called "hypercube" or "cube")
- Scorecarding, dashboarding, and data visualization
- Data warehouses
- Document warehouses
- Text mining
- DM—Data Mining
- BPM—Business Performance Management
- EIS—Executive Information Systems
- DSS—Decision Support Systems
- MIS—Management Information Systems

DESIGNING AND IMPLEMENTING BPM

When implementing a BPM program one might like to pose a number of questions and make a number of resultant decisions, such as:

- *Goal alignment queries:* The first step is determining what the short- and medium-term purpose of the program will be. What strategic goal(s) of the organization will be addressed by the program? To what organizational mission or vision does it relate? A hypothesis needs to be crafted that details how this initiative will eventually improve results and performance (i.e., a strategy map).
- *Baseline queries:* Current information gathering competency needs to be assessed. Do we have the capability to monitor important sources of information? What data is being collected and how is it being stored? What are the statistical parameters of this data; for example, how much random variation does it contain? Is this being measured?
- *Cost and risk queries:* The financial consequences of a new BI initiative should be estimated. It is necessary to assess the cost of the present operations and the increase in costs associated with the BPM initiative. What is the risk that the initiative will fail? This risk assessment should be converted into a financial metric and included in the planning.
- *Customer and stakeholder queries:* Determine who will benefit from the initiative and who will pay. Who has a stake in the current procedure? What kinds of customers/stakeholders will benefit directly from this initiative? Who will benefit indirectly? What are the quantitative/qualitative benefits? Is the specified initiative the

best way to increase satisfaction for all kinds of customers, or is there a better way? How will customers' benefits be monitored? What about employees, shareholders, distribution channel members?

■ *Metrics-related queries:* These information requirements must be operationalized into clearly defined metrics. One must decide what metrics to use for each piece of information being gathered. Are these the best metrics? How do we know that? How many metrics need to be tracked? If this is a large number (it usually is), what kind of system can be used to track them? Are the metrics standardized, so that they can be benchmarked against performance in other organizations? What are the industry standard metrics available?

■ *Measurement methodology-related queries:* One should establish a methodology or a procedure to determine the best (or acceptable) way of measuring the required metrics. What methods will be used, and how frequently will data be collected? Are there any industry standards for this? Is this the best way to do the measurements? How do we know that?

■ *Results-related queries:* The BPM program should be monitored to ensure that objectives are being met. Adjustments in the program may be necessary. The program should be tested for accuracy, reliability, and validity. How can we demonstrate that the BI initiative, and not something else, contributed to a change in results? How much of the change was probably random?

Need for BPM

Since the invention of double-entry bookkeeping, financial management and management by numbers continue to increase in importance to such an extent that no business today, whether small or large, makes decisions without financial data in hand.

This need for financial management gave rise to management–decision support systems driven by repositories of financial information. The repositories evolved into central warehouses of historical information or data warehouses. Managers quickly realized that the data warehouse was useful for more than just financial information; it could be used to better understand customers, to enhance marketing and merchandising, and to optimize their supply chain.

Business process integration arose from the need to connect applications. Application APIs evolved into Remote Procedural Calls (RPCs) between applications, which gave rise to Enterprise Application Integration (EAI). EAI extended the point-to-point connection with an integration hub that provided transformations, routing, and coordination. It quickly became clear that the value of this external integration node extended beyond application integration: it could also be used for externalization of processing from the applications.

As a result, business processes are no longer locked inside individual business applications. By moving the business process outside each application and associating it with a set of modeling and process choreography tools, new levels of flexibility were introduced into the business environment. If the business process needed to be modified or extended, it was no longer a code development effort.

This flexibility meant businesses were presented with new opportunity. The growth of business intelligence provided information about business performance. And BPI provided a vehicle for process reengineering to improve business operations. If these two attributes could be combined, a more intelligent and flexible business process environment could be created.

Achieving Performance Improvement

Process improvement is a structured methodology that involves leadership commitment, objective setting, organizational empowerment, and accountability at all levels. Its success requires a thorough knowledge of the methodologies and tools of process improvement, along with the experience to apply them to each unique process target.

The process and process steps are characterized, the measures are in place to drive results, and the goals seem almost unattainable. What stands between today's performance and the goals are obstacles. These obstacles must be removed to enable processes to operate at the target level. To remove obstacles effectively requires a disciplined methodical approach, outlined as follows.

1. From the measures, identify the gap existing between current performance of the process and the goal. This is the opportunity gap. As obstacles are removed from the process, the gap will close.
2. Identify the obstacles in the process that prevent it from operating at the goal.
3. Prioritize the list of obstacles to be removed. It is important not to try and fix everything at once. Work on those that have the most impact on the measure and are easiest to remove first.
4. Assign a small team of those most familiar with the issue to remove the obstacle. Clearly communicate what the obstacle is, the scope of the team's responsibilities, and the time allowed to complete the task.
5. Monitor the performance of this small team to ensure that they are making progress.
6. Subsequent to removal of the obstacle, review the measure to see if the removal action improved the measure and progressed toward the goal.
7. Repeat steps one through six until all obstacles are removed, and the process is at the desired performance level.

One thing to note in this approach is that even though obstacles are being removed, measures may not show any improvement. This can result from removing obstacles that were not truly obstacles in the process or from having the wrong measure. The correction of this problem requires a reassessment of the measure and the prioritized obstacles.

There is one more issue to consider relative to obstacles. In the workplace, ways to accomplish tasks will often be found in spite of the obstacles. To do this, processes and systems that circumvent the established processes are installed because they work when the processes do not. However, processes that circumvent obstacles must also be removed. In the optimized process they are now no value-added efforts.

THREE WAVES OF PERFORMANCE IMPROVEMENT

Enterprises have looked to exploit technology to maximize performance and competitive advantage in several distinct, although intricately overlapped, waves: through operational efficiency, decision support models and systems, and event-driven BI.

First Wave: Operational Efficiency

For competitiveness and profitability, enterprises have used technology to reduce the cost of business processes and procedures, and to improve customer service. Innovative organizations have implemented such systems as workflow management, front office CRM, and ERP (Enterprise Resource Planning) to accomplish these goals.

These systems generate and collect enormous amounts of data. In order to maximize operational efficiencies, organizations needed to use that data to gain insights into their business.

Second Wave: Decision Support Models and Systems

Businesses have always wanted to use technology to analyze, predict, and plan with clarity and insight. To this end, they have implemented business intelligence, data warehousing, advanced planning, and scheduling solutions.

However, business and the marketplace are dynamic and few organizations are agile enough to respond effectively to the transactional events altering the situation on a moment-by-moment basis. The key to maintaining and increasing competitiveness in today's high-speed market is time, hence, the need for event-driven business intelligence.

Third Wave: Event-Driven BI

Effective business performance management is the driving force behind the third wave. Decision makers must have a way to measure and manage their business processes. Key performance indicators, scorecards, and dashboards give managers meaningful metrics to monitor the performance of these processes.

As the ability to adapt quickly to change becomes a fundamental requirement and as the time to react to business events is squeezed by competitive pressures, corporations are recognizing the crucial importance of event-driven BI as the next wave of innovation.

Event-Driven BI

Organizations know that their bottom line can depend on acting on information within minutes rather than hours or days. They also know that information can be both a blessing and a curse. As organizations accumulate more and more data, it takes more and more time for users to sift through it in order to find the information they need to take action. The time between an event and the resulting action must be compressed.

Event-driven business intelligence compresses this time gap by "pushing out" time-sensitive information to users (or "subscribers"), dramatically increasing the speed with which they can access such information. Essentially, event-driven BI monitors three classes of events in operational and business intelligence content—notification, performance, and operational events—looking for key changes.

Having detected changes, event-driven BI then notifies and alerts decision makers, keeping them informed and up to the minute. This personalized information can be pushed to decision makers no matter where they are, enabling them to make timely and effective decisions.

With event-driven BI, a business can begin to manage its risks and opportunities as close to when the business events occur as possible. Critical information has to find the decision maker quickly, wherever he or she is located, so that further action can be taken in a timely manner. The second factor driving the adoption of event-driven business intelligence is the changing way in which we work and do business. As the time between the occurrence of events and the need to take action is compressed, individuals must be able to react more quickly. Changing work styles and paradigms are driving the adoption of event-driven BI. The individual worker is forced to make decisions quickly and to share relevant information across departments that will affect the overall competitiveness of the business. This change is a crucial factor driving the adoption of event-driven business intelligence.

A third significant factor driving the adoption of event-driven BI is the worker's relationship to information. Today's decision makers are drowning in information. They want the right information, at the right time. And, when it comes to important information, they would like such information to find them. In general, the decision maker seeks three types of data:

1. *Ongoing information*: Data that the business person wants to keep track of on a regular on-going basis
2. *Deep data*: Information that gives an in-depth view of the business or business process and which can be utilized to make insightful business decisions
3. *Time-critical data*: Information that must be received quickly and on a need-to-know basis because it could have a critical impact on both foreseen and unforeseen risks and opportunities

Zero Latency Enterprise

Although all three layers of information are crucial to overall business success and ongoing competitiveness, time-critical data is quickly growing in importance. As a result of this increased importance, Business Activity Monitoring (BAM) is used to describe the information management layer needed to manage and control the zero latency enterprise. The zero latency enterprise, or real-time enterprise as it is sometimes known, refers to an enterprise in which data generated by activities taking place in one part of the organization is immediately made available to workers in any other part, wherever they are working. The zero latency enterprise requires zero latent management and BAM is the solution that provides management information in real or near real-time.

Enterprises that want to make their business more effective and competitive must begin to incorporate event-driven BI in their overall BI framework and move the time horizons or information latency to the ever-shorter time periods of BAM. Throughout the world, in every industry, enterprises—and, most important, individuals within those enterprises—want to keep their finger on the pulse of the business and the market.

IMPACT OF BI AND BPM ON THE TELECOMMUNICATIONS INDUSTRY

High-performance BI tools such as advanced analytics, reports, and executive dashboards are playing a vital role in the telecommunications industry today. The value for the enterprise lies in the challenge of aggressively retaining and growing the customer base and extracting more value from each customer relationship. Those departments receiving the greatest benefit from BI and BPM include:

■ *Marketing:* To oversee product marketing and product management activities
■ *Customer care:* To improve customer retention and customer satisfaction levels
■ *Call center:* To better manage up-sell, cross-sell, and outbound marketing campaigns
■ *Network operations:* To retain and improve network and service quality management
■ *Finance:* To monitor cost per gross add, marketing spending, plus the impact of churn

BI and BPM tools enhance core functionalities such as customer care, sales, billing, and service delivery, as they:

■ Provide analysis and visibility into business metrics for decision makers
■ Facilitate improvement of operational processes
■ Support one integrated view of the customer across all departments, processes, and products
■ Give enterprisewide visibility of revenue, cost sources, and allocation
■ Increase revenue from greater understanding of business processes and customer segments

With advanced BI tools and BPM telecommunications companies are today addressing the key areas of customer insight and revenue growth including the following.

Average Revenue per User (ARPU) Boosting

ARPU boosting provides operators with the capability to identify and deploy potential revenue-increasing programs within the customer base. BI and BPM tools help operators in the improvement of ARPU and revenue stream analysis, the ability to identify segments with propensity to buy, market penetration analysis, and campaign effectiveness analysis.

In most mature markets, communications service providers are experiencing a decline in fixed access lines and are near saturation in terms of mobile customer penetration. Due to diminishing subscriber growth, network operators focus on retaining profitable customers as well as driving incremental revenue from the installed base.

Churn Prediction and Management

It is six times more costly to acquire a new customer than to retain a profitable customer. Churn prediction and management provide communications service providers with the means to more effectively retain their most valuable customers, segment their customer base, and focus marketing campaigns in terms of Customer Lifetime Value (CLV).

Understanding customers can have a positive impact on business. BI and BPM tools can help create transparency in your customer base by identifying potential churners, determining customer lifetime value, evaluating effective retention programs, and identifying the most responsive targets.

Integrate Cost and Revenue

The cost of customer acquisition and customer profitability is pushing companies to look more closely at profit margin analysis to reduce costs and increase revenue. The expansion into multi-service, multi-network, and multi-partner environments has made cost management complex, because many carriers don't truly understand the accuracy of inventory and assets, customer profiles, and business users' needs.

While carriers struggle to compete by offering creative service bundles, they often fail to conduct detailed margin analysis to understand the costs associated with those bundles, making it difficult to implement profitable plans and eliminate unprofitable services, and customers. For carriers to understand their profit margins there must be a clear view of both costs and revenues on a per-product and per-customer basis. This requires that carriers change their mindset in treating Revenue Assurance (RA) and Cost Management (CM) as two totally different departments within their business.

Ultimately, cost management departments will have to communicate and share data with revenue assurance departments so that marketing can truly understand the impact of acquisition costs on profitability. This will require using fine-grained revenue measures that exceed the capability of existing general ledger and accounting systems. Service provider profitability will increase as the result of the RA groups' focus on identifying revenue leaks, and the cost management groups' managing and reducing costs. For synergies to be realized and control margins to improve, there need to be end-to-end revenue monitoring systems that start with provisioning of services and continue all the way through to invoice verification in cost management. In other words, if carriers look at only what they bill, without marrying information to what they know of costs (i.e., what wholesale providers charge), then they have only half the picture.

Because most companies are not at that point of sophistication to truly integrate cost and revenue management, some are managing to derive business intelligence directly from billing and OSSs via rapid-query tools, analyses, and reporting capabilities. As BI and BPM platforms become more prevalent in telecom, they will help carriers to establish integration among financial measures of profitability and operational key performance indicators.

Improved Procurement

Most companies can have only a rough idea of the productivity gains that can be derived from SCM (Supply Chain Management) as part of the entire setup. No manufacturing company can get a 360-degree view of its operations if its SCM is a stand-alone system. BI has also made inroads in providing a consolidated view of the company including its supply chain.

This has helped reduce the traditional limited view of the supply side of operations. With this new approach of BPM, the BI system also acts as an intermediary between the organization's CRM and SCM systems. The consolidated view aside, this approach helps the company manage its supply to match the demands of its customers. It can also help the company tailor product attributes to match changing customer needs.

Self-Service

Economic success and prosperity for businesses in the twentieth century owe their origin to customer empowerment brought about by one over-riding development: the introduction of self-service in almost every business area. The deployment of next generation services, such as broadband, VoIP, IPTV, wireless TV, and FTTH will create customer pressure on telecommunications service providers to provide for customers' self-control of their own services and support. Self-service enables service providers to not only reduce the cost of interaction, but also collect more customer information to enable more personalized service. This, in turn, can drive higher customer retention and increase revenue. From the customer's perspective, self-service is valuable because it is convenient and flexible. A well-designed self-service system can help companies not only reduce customer care costs, but also help retain customers and provide up-selling and cross-selling opportunities.

Although self-service technology is now widely taken for granted, it has transformed the way business operates, with business intelligence at the heart of any successful self-service initiative. Business intelligence tools improve overall customer satisfaction by making self-services more inter-active and productive by making available and using information from multiple sources. It also makes such initiatives more relevant by applying experience and assumptions to develop an accurate understanding of customer usage dynamics.

Business intelligence tools enable service providers achieve optimal effectiveness, while investing in self-service initiatives by assisting them in:

- Collecting sufficient, current, and accurate data
- Ensuring information and content are up to date and effective
- Integrating front-end and back-end systems
- Providing customers features that enable control, customization, and ease-of-use
- Ensuring flexibility for service providers for offering new services and promotions

Risk Resolution and Management

Risk management is increasingly viewed as an integral element of the risk and decision making by organizations. Risk management may be undertaken prior to the decision itself (e.g., insuring against certain risks) or after the decision (i.e., effective management of relationships with customers to reduce the likely incidence of disputes).

These uncertainties and risks place new demands on the field of risk management, indicating that the conventional methods of managing uncertainty (e.g., buffer stocks, spare capacity, quoted lead times) are likely to be less effective in meeting the new demands and uncertainties.

With BPM and BI tools managers are now equipped with improved knowledge, skills, and understanding to be able to identify, analyze, and manage these developments and to assess the consequences and risks arising from the more diverse range of market contexts. BI provides the necessary resources and solutions to facilitate this improvement in risk management.

Pre- and Postpaid Convergence

Traditional segmentation of wireless subscribers based on their payment modes has resulted in separate OSS/BSS systems. Fierce global competition is now challenging the basis of this segmentation. The economies of maintaining two separate systems, the need to introduce advanced services and pressure to reduce TCO (Total Cost of Ownership) is compelling service providers to redraw the OSS/BSS solutions.

Subscribers across the world markets have selected prepaid as a preferred payment method over postpaid. At the same time they demand the products and services offered to other postpaid customers. Existing prepaid systems heavily rely on the network-based charging, which can deal with time-based charging but are incapable of charging the combination of voice, data, and content services.

For service providers the additional revenue streams that can be generated through offering advanced services to the postpaid subscribers is limited by the ability to charge for these services. With pressures to offer competitive services and increase ARPU, service providers find the ever-increasing need to adopt a converged BSS solution for their prepaid and postpaid customers.

Prepaid services, primarily introduced to develop new market segments, meant the service providers had to add stand-alone systems. These systems are limited in terms of scalability, rapid growth in the subscriber base, flexibility, and complex offerings of advanced services. Efforts to migrate prepaid subscribers to postpaid services to overcome these challenges have not met with the desired results.

This has led service providers to think of alternate strategies and the emerging need for a converged prepaid–postpaid solution to handle all the rating, billing, and customer relationship management demands. The converged solution provides service providers with operational efficiencies, increased customer satisfaction, and the ability to introduce products and services quickly and efficiently to subscribers regardless of their preference of payment method. With all the capabilities of their postpaid systems available to prepaid subscribers, the difference between these types of subscribers is reduced to a choice of payment method.

Business Challenges

Hybrid Prepaid and Postpaid Accounts: Service providers cannot offer bundled prepaid and postpaid services and price plans.

Advanced Products and Services: Many complex voice and data services cannot be offered to prepaid subscribers as the prepaid systems cannot charge them.

Integrated Customer View: Individual subscriber needs cannot be addressed.

Customer Service: Limited communication to prepaid subscriber restricting target promotions.

Operational Challenges

Vendor: Different vendors for prepaid and postpaid systems result in integration, upgrade, and maintenance issues.

IT and Operations: Separate database, operations teams associated with prepaid and postpaid resulting in overheads and reduced bottom lines.

Customer and Product Data: Prepaid and postpaid customer, product data maintained separately resulting in duplication and increased cost of maintenance.

OSS/BSS Infrastructure: Diverse infrastructure resulting in higher maintenance costs and integration challenges.

The business technical and operational challenges and inefficiencies resulting from disparate prepaid and postpaid billing systems if addressed effectively can offer increased revenue streams and give a much required boost to the service providers' bottom line.

BI for Convergence

Currently, most service providers are maintaining separate systems for prepaid and postpaid. This situation has many drawbacks on both the marketing and operational levels, including duplication of effort, and the lack of an integrated view of the customer. The end result is that service providers have higher operational costs and yet are not able to generate new revenue generation streams.

There are huge benefits of moving to a convergent BI environment which includes a single set of system modules, common infrastructure, and rapid introduction of new services. However, flexibility offered by the BI platform in transition from the current separate prepaid and postpaid systems to a convergent system is unprecedented. Most BI platforms allow transition in several ways, in accordance with the customer's requirements. The approach can be a rapid transfer to a single convergent billing system, or phased, in a manner that guarantees benefits and ROI at each stage. BI provides a smooth migration path from legacy systems.

Such platforms enable the service provider to nurture its prepaid customers through a unified customer view, sophisticated offerings for next generation services, and enhanced customer service. Moreover, it can offer the benefits of maintaining a single billing system for prepaid and postpaid, with a single customer database, product catalog, and rating engine. This approach is effective across multiple functional areas and provides more efficient applications and services such as:

- Message acquisition and formatting
- Real-time charging environment
- Multidimensional rating
- Balance management
- Product catalog, development, and improvement
- Single customer data and billing

9

BUSINESS INTELLIGENCE: ISSUES AND CHALLENGES

The concept of Business Intelligence (BI) runs the risk of ending up in the "failed to meet expectations" heap along with technologies and solutions such as Customer Relationship Management (CRM) and Enterprise Resource Planning (ERP).

The promises of BI—that people in the organization have access to information and use that information to improve the business—has largely not proven out. Business intelligence has not lived up to its hype, missing the mark in at least two areas: broad distribution and timely informed decisions.

Business intelligence encompasses tools for data Extraction, Transformation, and Loading (ETL), data warehouses, data marts, data cubes, data mining, reporting tools, ad hoc query tools, and industry solutions, however, few vendors of business intelligence provide all the functionality encompassed in this definition. More important, few vendors provide what end customers really need: simple to use, easy to manage solutions that provide useful insight into information for all decision makers. Regardless of the technology or the definition of the solution, users want and need access to enterprise information that helps them make timely informed decisions.

Based on the findings of various surveys of business intelligence and OLAP (OnLine Analytical Processing) users, many business intelligence solutions have been deployed to only a handful of users in an organization, and those that have used the solutions have found them difficult to use and understand. Too much data is still locked up in transactional systems and not used to make decisions. Too many users are still waiting for a report to have the information they need to improve business processes. Those users who do have business intelligence applications—reporting and ad hoc query tools mostly—can see only a slice of the enterprise data, so their view of the data is occasionally deep, but rarely broad.

Most BI solutions are deployed to only a limited group of users rather than to decision makers throughout the company. Additionally, BI often fails to provide actionable information to the people who need it. For BI to have an impact on the way decision makers work, it must provide information to a wide range of users, be easy to use and understand, and help people make decisions in real-time about their business.

THE BI GHETTO

A typical BI project has an average four-time Return On Investment (ROI), but due to some challenges and issues, organizations are unable to fully benefit from a global, cross-functional analysis of information. A significant disappointment has been the failure of BI to involve more than five percent of the enterprise in making data into intelligent and actionable information. BI represents the exclusive analytic realm where only the most highly trained data moguls ("power users" inside and outside IT) spend their most productive time. This has become the true legacy of BI, and not one that should be applauded, but rather one that must be torn down.

BI today faces a situation that some industry pundits have described as "the BI ghetto." The reasons behind BI's current disappointments are that BI tools and products have been licensed toward individual users, inherently restricting broad deployment whereas they have been targeted at a select group of users within the enterprise, including database administrators, data analysts, and application developers, the sum of whom is no more than five percent of an enterprise's workforce.

Furthermore, business has pushed BI to the corner by analytic complexity and application logic and IT has not taken an active role in developing a BI strategy, nor were they encouraged to actively participate by either the vendors or the business units they support.

CRITICAL CHALLENGES FOR BUSINESS INTELLIGENCE SUCCESS

More than half of all business intelligence projects are either never completed or fail to deliver the features and benefits that are optimistically agreed on at their outset. Although there are many reasons for this high failure rate, the biggest is that companies treat BI projects as just another IT project.

The reality is that business intelligence is neither a product nor a system. It is, rather, a constantly evolving strategy, vision, and architecture that continuously seeks to align an organization's operations and direction with its strategic business goals.

With BI, business success is realized through rapid easy access to actionable information. This access, in turn, is best achieved through timely and accurate insight into business conditions and customers, finances, and markets. BI is defiantly complex, but the returns it provides make it worth the effort. Successful BI brings greater profitability, the true indicator of business success. And success is never an accident; companies achieve it when they do the following:

- Make better decisions with greater speed and confidence.
- Streamline operations.
- Shorten their product development cycles.
- Maximize value from existing product lines and anticipate new opportunities.
- Create better, more focused marketing as well as improved relationships with customers and suppliers alike.

Organizations must understand and address challenges critical for BI success. BI projects fail because of:

- Failure to recognize BI projects as cross-organizational business initiatives, and to understand that as such they differ from typical standalone solutions
- Unengaged business sponsors (or sponsors who enjoy little or no authority in the enterprise)
- Unavailable or unwilling business representatives
- Lack of skilled and available staff, or suboptimal staff utilization
- No software release concept (no iterative development method)
- No work breakdown structure (no methodology)
- No business analysis or standardization activities
- No appreciation of the impact of dirty data on business profitability
- No understanding of the necessity for and the use of metadata
- Too much reliance on disparate methods and tools (the dreaded quick pill syndrome)

In this section, we examine each of these challenges.

Cross-Organizational Collaboration

Traditionally, any business initiative, including a decision support project, was focused on a specific goal that was limited to a set of products or an area of the business. Due to this narrow focus, organizations were unable to analyze the project's impact on business operations as a whole. As organizations became more customer-focused, these initiatives began to integrate customer information with product information. It is critical to realize that customers and markets, not manufacturing plants and product managers, must drive the business. It is also optimal to correct any customer problems before the customer realizes the problem exists. Enterprises have a better chance to achieve high customer loyalty if customers can pay when their problem is solved, not when the product is shipped.

Initially, the integration occurred in regional or departmental databases, with no cross-regional collaboration. Enterprise data warehouses were the next step in the evolution towards cross-organizational integration of information for decision support purposes such as sales reporting, Key Performance Indicators (KPIs), and trends analysis.

Customer relationship management followed, bringing the promise of increased sales and profitability through personalization and customization. BI is the next step in achieving the holistic cross-organizational view. It has the potential to deliver enormous payback, but demands unprecedented collaboration. Where BI is concerned, collaboration is not limited to departments within the organization; it requires integration of knowledge about customers, competition, market conditions, vendors, partners, products, and employees at all levels.

To succeed at BI, an enterprise must nurture a cross-organizational collaborative culture in which everyone grasps and works toward the strategic vision.

Business Sponsors

Strong business sponsors truly believe in the value of the BI project. They champion it by removing political roadblocks. Without a supportive and committed business sponsor, a BI project struggles for support within an organization, and usually fails. Business sponsors establish proper objectives for the BI application, ensuring that they support the strategic vision. Sponsors also approve the business-case assessment and help set the project scope. If the scope is too large, sponsors prioritize the deliverables. Specifically for BI projects, business sponsors should also launch a data-quality campaign in affected departments. This task goes to business sponsors because it's business users who truly understand the data. Finally, business sponsors should run a project review session at assigned checkpoints to ensure that BI application functionality maps correctly to strategic business goals, and that its ROI can be objectively measured.

Dedicated Business Representation

More often than not, the primary focus of BI projects is technical rather than business-oriented. The reason for this shortcoming: most BI projects are run by IT project managers with minimal business knowledge. These managers tend not to involve business communities. Therefore, it's not surprising that most projects fail to deliver expected business benefits.

It's important to note that usually 20 percent of the key business people use BI applications 80 percent of the time. Therefore, it's vital to identify key business and technical representatives at the beginning of a BI project, and to keep them motivated throughout the project.

A BI project team should have involved stakeholders from the following areas:

Business executives are the visionaries with the most current organizational strategies. They should help make key project decisions and must be solicited for determining the project's direction at various stages.

Customers can help identify the final goals of the BI system. After all, their acceptance of products or service strategies is what matters most.

Key business partners provide a different view of the customer and should be solicited for information at the start and on an ongoing basis.

The finance department is responsible for accounting and can provide great insight into an organization's efficiencies and improvement areas.

Marketing personnel should be involved during all phases of the project because, typically, they are key users of BI applications.

Sales and customer support representatives have direct customer contact and provide customer perspective during a BI project. They must have representation on the team.

IT supports the operational systems and provides awareness about the backlog of BI requests from different groups. In addition to providing technical expertise, the IT staff in the BI project team must analyze and present BI-related requests.

Operations managers and staff make tactical business decisions. They provide the link between strategic and operational information, making them important during some key phases of a BI project.

Availability of Skilled Team Members

BI projects differ significantly from others because, at their outset, they tend to lack concrete, well-defined deliverables. In addition, the business and technical skills required to implement a BI application are quite different from other operational OnLine Transaction Processing (OLTP) projects. For example, operational projects normally focus on a certain area of the business, such as ERP, CRM, or Supply Chain Management (SCM); however, a BI project integrates, analyzes, and delivers information derived from almost every area of the business as a whole.

The required technical expertise varies as well; typically, for example, a database administrator's focus is efficient retrieval of data using OLTP systems. By contrast, where BI systems are concerned, it's vitally important to focus on data storage in addition to data retrieval. A BI project team lacking BI application implementation experience will most likely fail to deliver desired results in the first iteration. Because most BI projects have aggressive timelines and short delivery cycles, an inexperienced and unskilled team is a risk that must be avoided.

Mandatory BI project skills include:

BI business analysts who can perform cause-and-effect analysis to develop business process models for evaluating decision alternatives. These individuals should also be able to perform what-if analysis by following a proven BI methodology.

A KPI expert experienced in creating balanced scorecards. These experts must be able to identify the KPIs that meet business needs, calculate and report them, and monitor performance. They also should iteratively reevaluate KPI effectiveness and must integrate these KPIs into the balanced scorecard.

Balanced scorecard experts to continuously develop and fine-tune scorecards. Measuring success in a dynamic business environment requires an effective toolset. With a balanced scorecard, an organization's vision and strategy can be translated into objectives, targets, and metrics, and incentives to meet those objectives and targets.

Data warehouse architects with experience developing BI-related logical and physical data models, including both star schemas and OLAP. Ideally, these people might also have experience with such technologies as statistical tools and data-mining algorithms.

Cube developers and implementers with experience implementing BI-specific data models, OLAP servers, and queries. These individuals must be able to develop and deploy complex and intelligent cubes to conduct multidimensional OLAP analysis for different users.

Personalization experts experienced at developing Web-based generic BI applications that can not only meet the reporting needs of many users, but also provide a personalized view to each user.

BI APPLICATION DEVELOPMENT METHODOLOGY

To succeed, BI projects must adhere to a plan with clearly defined methodologies, objectives, and milestones. In this respect, they are hardly unique. However, unlike other undertakings, BI projects are not limited to a confined set of departmental requirements. Rather, their purpose is to provide cross-organizational applications. Therefore, BI methodologies and deliverables differ.

As with any project, BI starts out by answering some basic questions, such as:

- What will be delivered?
- What are the benefits and expected ROI?
- What is the total cost?
- When will it be delivered?
- Who will do it?

The answers collectively define the BI project as follows.

Project deliverables map goals to strategic business objectives. These deliverables should be measurable in business terms. For example, "In order to increase sales 30 percent, the sales data merged with prospect data must be available to sales teams within three days of month's end."

Project scope aligns deliverables with BI application deployment phases and timelines. Unlike traditional OLTP applications, the number of transactions the system will perform cannot measure BI project scope.

Transactions usually represent an organization's processes, which in turn represent functions. Because BI projects are data intensive and not function intensive, their scope must be measured by the data they will transform to the target BI databases, and how quickly this data can be available. This focus on data is necessary because almost 80 percent of the effort in a typical BI project is spent on data-related activities.

ROI for a BI project must be derivable from project deliverables. Project sponsors must measure the effectiveness of delivered BI applications after the completion of each phase to determine whether the project is delivering the promised ROI. If it isn't, improvements must be made.

Planning BI Projects

Due to their nature, BI projects tend to hit more unknowns than OLTP projects. Why? OLTP projects implement the processes of an organization, which in turn represent the functions. By contrast, BI projects are supposed to provide data, which will be transformed into information, which in turn is transformed into action. Therefore, BI project planning is not a one-time activity, but rather an iterative process in which resources, timelines, scope, deliverables, and plans are continuously adjusted.

Although it's an iterative process, the initial project plan must be created with as much detail as possible.

BI project planning activities include the following:

- *Determining project requirements:* As part of this activity, existing high-level data, functionality, and infrastructure requirements must be reviewed and revised to include more detail and remove ambiguity.
- *Determining the condition of source files and databases:* Before completing the project plan, operational data stores must be reviewed to account for any issues that may surface during the data-analysis phase.
- *Determining or revising cost estimates:* During this activity, the organization performs detailed analysis to determine purchase and maintenance cost estimates for hardware, software, network equipment, business analysts, IT staff members, implementation, training, and consultants.
- *Determining or revising risk assessment:* Enterprises must perform a detailed risk assessment in order to accurately determine and rank BI project risks (based on severity and the likelihood of their occurrence).
- *Identifying critical success factors:* Here an organization determines what conditions must exist in order for the project to succeed. Factors include supportive business sponsors, realistic timeframes, and the availability of resources.
- *Preparing the project charter:* This is a detailed memorandum of understanding that should be prepared by the project team and approved by the business sponsor and key business representatives.
- *Creating a high-level project plan:* These are detailed breakouts of tasks, resources, time lines, task dependencies, and resource dependencies mapped on a calendar.
- *Kicking off the project:* On completion of the plan, the project is kicked off in an orientation session at which all team members, business representatives, and the BI sponsor are present.

Business Analysis and Data Standardization

By now it's clear that BI projects are data intensive and that "data out" is as important as "data in." It's crucial that the source data be scrutinized. The age-old saying, "Garbage in, garbage out," still holds true. In most BI projects, business analysis issues are related to source data, which is scattered around the organization in disparate data stores and in a variety of formats. Some of the issues include the following.

Identifying Information Needs

Most business analysts have challenges when it comes to identifying business issues related to BI application objectives. They must evaluate how addressing these issues can help in obtaining answers to business questions such as, "Why is there a decrease in sales revenue in the fourth quarter in International Market?" Once the issues are identified, business analysts can easily determine related data requirements, and these requirements can in turn help identify data sources for the required information.

Data Merge and Standardization

The biggest challenge faced by every BI project is its team's ability to understand the scope, effort, and importance of making the required data available for knowledge workers. That data consists of fragments in disparate internal systems and must be merged into a common data warehouse, which is not a trivial task. Data requirements normally extend beyond internal sources, to private and external data. Therefore, data merge and standardization activities must be planned and started at the beginning of the BI project.

Impact of Dirty Data on Business Profitability

Inaccurate and inconsistent data costs enterprises millions. It's imperative to identify which data is important, and then find out how clean it is. Any dirty data must be identified, and a data-cleansing plan must be developed and implemented. The business objectives of any BI project should be tied to financial consequences such as lost revenue and reduced profit. The financial consequences are usually the result of a business problem related to inaccuracies in reports due to reliance on invalid, inaccurate, or inconsistent data. However, most BI projects fail to tie financial consequences to dirty data through monetary expressions (such as losing $25 million in quarterly revenue due to the enterprise's inability to up-sell). Even the best BI application will be worthless if driven by dirty data. Therefore, it is important for every BI project to employ knowledgeable business analysts who understand the meaning of source data and can ensure its quality.

Underestimating the data-cleansing process is one of the biggest reasons for BI failure. Inexperienced BI project managers often base their estimates on the number of technical data conversions required. Project managers also fail to take into account the overwhelming number of transformations required to enforce business data domain rules and business data integrity rules. For some large organizations with many old file structures, the ratio of a particular data transformation effort can be expected to be as high as 85 percent effort in data cleansing and only 15 percent in enforcing technical data conversion rules. Therefore, even if estimates appear realistic at the project's outset, you must factor in data-cleansing efforts. Note that full-time involvement from the right business representatives is mandatory for data-cleansing activity.

Importance of Metadata

Clean data is worthless to knowledge workers if they do not understand its context. Valid business data, unless tied to its meaning, is still meaningless. Therefore, it is imperative for all BI applications to consciously create and manage the meaning of each data element. This data about data is known as metadata, and its management is an essential activity in BI projects. Metadata describes an organization in terms of its business activities and the business objects on which they're performed. It helps transform business data into information. It is imperative for every BI environment. For example, what is profit? Does every businessperson have the same understanding of profit? Is there only one calculation for profit? If there are different interpretations of profit, are all interpretations legitimate? If there are multiple legitimate versions of profit, then multiple data elements must be created, each with its own unique name, definition, content rules, and relationships. All this information is metadata.

Metadata helps businesspeople navigate BI target databases and help IT manage BI applications. There are two types of metadata:

1. *Technical metadata:* provides information about BI applications and databases, and assists IT staff in managing these applications.
2. *Business metadata:* provides business users with information on data stored in BI applications and databases.

Both types are crucial to success and should be mapped to each other and stored in metadata repositories.

The Quick Pill Syndrome

There is neither a single technology nor a technique that will resolve all the challenges to reach the goal of a successful BI environment. That is to say, there is no quick pill. BI projects have an enormous scope and cover multiple environments and technologies.

A BI environnent comprises:

- A tool for extracting, transforming, and loading data from disparate source systems into the BI target data warehouse
- A data warehouse that stores historical and current business data, as well as an OLAP server that provides analytic services
- Front-end BI applications that are used to provide querying, reporting, and analytic functions to the organization's knowledge workers

In most organizations, these BI components are implemented in different phases and by project teams. Each team implements the product that meets most of its functional requirements. More tools create greater complexity and increased interoperability issues, and require more administration involvement. BI project teams must always consciously strive for the lowest possible number of tools. This will allow different BI activities to map to the same over-all roadmap.

Customer Pain Points

Some of the top customer pain points are the following.

Relying on Inaccurate and Outdated Information to Make Business Decisions

Business managers today have a wealth of up-to-date information available, yet they frequently complain about having to call the IT department every time they need accurate, real-time data. Information systems and reports are incomplete and difficult to decipher. In large enterprises, incompatible or redundant business reporting tools are often the norm, leading to interdepartmental miscommunication and confusion. In addition, the number of people needing access to information is skyrocketing as business leaders recognize the value of corporate process transparency.

Overwhelmed with Inefficient Manual Processes

Businesses today are less likely than ever to be confined to one office, one geographic area, or one country, and business processes are continually in flux as a result. Organizations are looking for ways to improve communication, reduce time to market, and improve customer service.

Unable to Effectively Leverage the Intellectual Capital in Your Enterprise

Explosive growth in the content that organizations produce, and government regulations about how that content must be treated, put demands on business leaders to make information more accessible to the people who need it.

Using Expensive, Proprietary Technology to Connect Disparate Applications on Incompatible Systems

The smaller the world gets, the more interwoven is your web of customers, suppliers, partners, and employees. Proprietary technology can be expensive and in a global economy, businesses need the flexibility of secure developing environments and tools that allow them to connect efficiently to virtually any application in the world.

CREATING A COST-EFFECTIVE ENTERPRISE FRIENDLY BI SOLUTION

BI must deliver solutions that are more useful to the stakeholders. These must be more expressive, allowing business people to develop their own models as easily as they can in a spreadsheet. And they must help business users solve end-to-end problems, closing the loop by tying analytics directly to action and vice versa.

Make It Current

The latency in data warehouses, with updates occurring overnight at best, and even more infrequently for the downstream data structures that users actually see, is unacceptable for the emerging hybrid applications discussed in the previous section. In addition, unattended analytical processes, where queries to the data warehouse are initiated by other systems, without human involvement, are increasingly relying on real-time or near-real-time data. The elaborate designs of today's data warehouse simply cannot accommodate these requirements. Sourcing transaction-level and even subtransaction-level data, at enormous volume and extremely low latency demands extreme computing power, driving the cost up when using proprietary platforms and software to unacceptably high levels. Alternatives are now available.

Using high-performance, scalable data warehouse appliances solves a major challenge in data warehousing. In order to provide adequate query performance on constrained hardware platforms, design techniques such as star schemas, aggregates, heavy indexing, and data marts are employed. Unfortunately, these designs are notoriously poor at being updated, except in batch.

Normalized schemas, optimized for transaction processing, are more appropriate for trickle-feeding real-time updates and catching messages from queues, EAI transactions, Web services, logs of running programs, and a host of other sources, but these schemas typically offer very poor query performance.

Make It Simple

Any question that can be posed against a data model should be resolvable within a reasonable time at a reasonable cost. The business user should be able to think broadly about business questions to be answered and then use a simple interface to turn these into queries. Clearly, controls are needed to deal with naïve queries that can cause degradation or complex queries that are so resource-intensive they must be isolated in order not to affect the other users of the system, but this should be the exception, not the rule. Moreover, simplicity should not be limited to the business users of the system. Managing and enhancing the physical elements should also be simple. Adding new CPU, memory and storage capabilities, and performing load balancing, in short, all of the administrative and systems functions should be just as simple as posing a question to the system.

Relationships, mappings, and models are all conceptual. In any organization, they change constantly. The best set of models in a relational database, with views, indices, demoralizations, and partitioning, is only a good solution for a period of time. Unfortunately, organizations are often unable to apply the same level of discipline to the maintenance of these structures as they did to the original development effort, particularly given the multi-departmental and cross-functional nature of BI. Simplifying the structure by replacing physical optimization with more computing power makes a compelling case. Another compelling case is to employ software that manages the physical optimization of the data warehouse, by monitoring the usage patterns and constantly reconfiguring the physical layer without effort or incident above it.

A BI system must be easy to update in near-real-time to reflect real-time changes in the business environment. Unless businesspeople are provided with a BI solution that is at least as easy to use as a spreadsheet, they will continue to shun it. These are the minimal requirements:

- Declarative modeling (no programming or scripts)
- Intelligence about the data resources (implies active metadata)
- Collaboration capabilities (the ability to share work and distribute it)
- Complete abstraction from physical data such as tables, columns, files, and directories
- Zero impact from changes in the physical model
- Ability to build models from scratch, assemble models from components, and be assured of referential integrity over time

Make It More Expressive

Relational database models are not business models. They are business data models. They lack the essential depth and richness that are expected by businesspeople, such as the sequential logic found in financial statements. The actual logic for calculating profit and loss, balance sheet, and cash flow statements involves certain items in sequence, making decisions, and sometimes even solving simultaneous equations, all of which can be handled easily in a spreadsheet. Another type of analysis involves separating things into buckets, where the discriminant for bucketing is an attribute of the lowest level of detail available, such as action on a claim. Describing the logic for a model using this approach is very simple, but nearly impossible in a relational data model.

BI developed along two parallel paths that are still evident. Some tools are aligned with relational databases and are SQL-based. Their conceptual models range from simple row–column manipulations to very expressive analytical tools, but in the latter case, a very powerful SQL-generating engine is used to mask the complexity and difficulty of conducting an analysis against relational tables. Other tools developed to allow more comprehensive conceptual models, but use proprietary languages and data management approaches, such as multidimensional databases.

It is an open question whether all of these solutions can coexist under the umbrella of a unified conceptual model with a conceptual-to-physical interpreter. One thing is clear: taking advantage of high-performance resources in today's market will power these decisions.

10

BUSINESS INTELLIGENCE: STRATEGY AND ROADMAP

With the proper Business Intelligence (BI) software in place, you can begin considering all the core elements of your business as a single integrated whole.

Gaining a competitive advantage in the market used to be a straightforward, common-sense process. Enterprises could provide industry-leading products and services, setting the standard in their market. Or they could excel at marketing and sales functions and create an irresistible "buzz" within their prospect base.

Times have drastically changed. Today, gaining a competitive edge is extremely difficult. Enterprises that thrive in the future will be those that can act rapidly on information. It is all about seeing business information in a new way.

The problem is not a lack of information. The problem most businesses face today is that they don't use their information to their best advantage. The problem lies in the way businesses view and act on their data. Information needs have changed, from being able to collect static data only to now collecting business events, with new competitive pressures and governmental regulations. Furthermore, businesses need to reduce the timeframe from the time they collect information to the time they act on it. Failing to act on business events within a short timeframe can have serious consequences. If you miss too many opportunities, churn rates will spiral and revenues will plummet.

The telecommunications industry has certainly seen better days as the economy and hence the business expanded substantially in the last decade. Technology budgets were larger during the growth phase and fiscal optimism fostered independence among company departments, groups, and divisions seeking changes and encouraged technology purchases to promote growth. Acquire the best-of-breed and grow-at-any-cost attitudes prevailed during this exciting time.

In the present sober business climate and consolidation phase, renewed emphasis on corporate accountability has prompted an assessment of all big-ticket technology investments and purchases. Today, businesses cannot afford independent expenditures that do not add value in a coordinated fashion for the whole enterprise.

Organizations have come to realize that many of the same business applications and tools exist throughout the enterprise but produce little or nothing of value together, except information silos. The maintenance and licensing fees on each such IT tool are significant and to avoid such organizational resource waste enterprises must stop this fragmented proliferation of IT tools and better ensure a more consistent and manageable IT environment.

Now that fragmented proliferation has slowed down, businesses can develop a plan to salvage as many investments as possible, and devise a strategy to integrate the disparate technologies to move the enterprise forward. For business to take full advantage of the rising new technologies that are driving the IT industry today—including Operational Data Stores (ODSs), Data Warehouses (DWs), BI, enterprise portals, Extraction, Transformation, and Load (ETL) tools, Enterprise Application Integration (EAI), and Web services—organizations must work toward converging those technologies. The new technologies must function symbiotically, coalescing critical information from across the organization and delivering it through a single access point to provide a collaborative environment for decision making.

Whatever the political or business reasons, the fates have not been kind to telecom over the past several years. In order to turn their fortunes, many telecom companies are investing heavily in new technologies that will enable them to collect and make sense of the overabundance of data that resides in their organizations, thus turning that data into usable information.

Business intelligence has seen much investment in the recent past, and is central to this problem of duplication and resource bleed. However, over the past decade or so, countless vendors have touted their products as all-inclusive or comprehensive solutions to DW/BI problems. As a result, many organizations have sunk copious amounts of money into these panaceas. They have built specialized data warehouses, operational data stores, data marts, and business intelligence applications on top of these data stores or legacy systems.

The pieces—DWs, BI tools, enterprise portals, ETL, and EAI—are already in place at many organizations. The key is to integrate those pieces, to converge them. The need to integrate ETL, EAI, DW, enterprise portals, and BI is not only important but is vital for the survival of the battered telecommunications industry where the use of multiple tools is the standard. The challenge, therefore, becomes how to seamlessly integrate a wide variety of BI tools and maintain the technology as well as the environment, posing a huge challenge to a fragmented BI/enterprise portal integration.

In order to survive and thrive, telecommunication companies need to integrate their BI capabilities to form a collaborative business platform that provides accurate, up-to-date, real-time, usable information.

PLANNING A BUSINESS INTELLIGENCE SOLUTION

An organization having a successful business intelligence implementation has distinct advantages over its market rivals. What it knows about its competitors, markets, customers, products, and operations allows it to substantially increase revenues, reduce costs, and enhance profitability. However, as the BI solutions become more pervasive and complex so does the cost associated with them. Not just the cost of the solution or the deployment cost, with the unprecedented proliferation of BI technologies across the fabric of most present-day top organizations the stakes are much higher. The impact of BI initiative failure can be hard hitting and in a few cases fatal.

On the flip side, it's easy to measure the hard dollar costs of operating an environment of patchwork applications that duplicate staff, hardware, software, and maintenance. However, it's the potentially huge and hard to measure cost of a delayed or wrong decision that could have the largest impact on bottom-line results. Most companies can live with a one-off event, but if your competitors have put the processes and systems in place to consistently make better decisions, then you are at an extreme disadvantage.

This suggests the need for a practical but all inclusive assessment process to determine necessities and expectations from a BI solution, careful analysis, and cost projections to ensure IT resources aren't overwhelmed by burgeoning use of the tools. Organizations must avoid rollout BI solutions in an iterative fashion and must give adequate consideration to the optimum way to deploy, upgrade, and maintain their software assets. Also worth considering is the development of an organizational roadmap articulating how BI technologies are used today and how they will be used into the future.

In short, organizations must first evaluate whether to go for BI implementation.

To Go or No-Go

The BI challenge is governed by several factors and all need to be given their due share of attention before a BI go/no-go decision and perhaps a BI roadmap evolves. Foremost among them are the need to set expectations right, knowing clearly what the business wants to use BI for, evaluating and understanding an organization's culture, respect for data quality, and, more significantly, availability of such data in mature, broadly integrated systems.

Understand Limitations of BI

BI is not a silver bullet solution to all decision-making problems. BI will provide an organization with the right tools and capabilities that will reveal to them underlying hidden patterns, answers not very obvious, and will also display the information in graphical format. However, BI cannot analyze on its own; organizations will have to devise a structure that will let them "slice and dice" the data and analyze it. Moreover, BI on its own cannot determine what to analyze either. The user is the driver, he sets out on a mission to come up with an analysis, knowing very clearly what to analyze. It is therefore important to evaluate up front what kind of analytical capabilities the BI solution will be expected to support or handle.

BI Usage

BI's strength comes from its analytical reporting capabilities that enable business users to slice and dice data, and go on an exploration mission to seek answers for the what-if analysis. BI is not really intended for operational level details, although almost every product supports it and carries drill-down features for further depth into operational data. Needless to say, substantial operational reporting continues to happen from BI tools. These users are mostly the lower levels of an organizational hierarchy using the BI platform as an alternate source of getting preaggregated transactional data to meet their routine reporting needs.

The analytical reporting usage of BI comes from the top 10 or less percent of an organization's users. These are business leaders and teams closely working with them who chase patterns, and mine for uncovered customer behaviors or market trends. Excessive usage of BI tools for operational reporting loads the BI systems with the burden of being in sync with transactional systems on a more frequent basis.

Organization Culture

Organizational culture controls whether its environment encourages users to challenge the use of data, and demand for information. There are many enterprises that run a report every month's end just because the report is a part of the standard operating procedure for month-end processing.

Does the business really know how many people actually use that report? In such an organization, the user community is by and large in an "execute" mode, carrying out functions day after day, month after month, the way they were laid down for it. Such an organization will have problems to come up with a correct BI strategy, much less successfully implement and sustain one. IT will look at business, and business will expect IT to tell it what shape and form the solution should take.

Is the Organization Ready for BI?

Organizations having many of the characteristics (symptoms) listed below must take adequate due-diligence measures before embarking on the BI path as the chances of disappointment are very high for such organizations.

- Enterprise systems don't exist; Excel is the system of choice.
- Users chase reports instead of information.
- More time is spent in fine-tuning reports instead of improving data.
- Standards are not implemented or not used with rigor.
- Systems of record are not stable or the concept does not exist.
- Some data is not captured at all, or IT does not know who holds that data.
- IT does not work with business users.
- BI solution appears to be fragmented.
- Organization is going to change soon: structural reorganization, merger, or acquisition.
- Either or both IT and business users do not believe that the organization will really benefit by implementing BI.
- Business leaders cannot champion the cause for BI and provide ongoing efforts to sustain the initiative.

BI is a very powerful tool but at the same time a difficult animal to tame. There are more organizations that have not reaped the benefits or meaningful return on investment than there are those enjoying market leadership and profitability through accomplished use of BI. Several organizations even years after the implementation still continue to use BI for operational reporting purposes. Unless your organization is really ready to consume and benefit from the analytical reporting muscle that BI touts, it is rushing toward a BI implementation that is slow on ROI (Return On Investment). A sound BI strategy always begins with an internal audit of a business to determine its preparedness and its capabilities to nurture and sustain a deployed solution.

Managing Total Cost of Ownership for Business Intelligence

Financial analysis of enterprise costs, such as technology investments and expenses, should be done regularly to assess the impact of changes that may affect the business. It is critical to quantify the Total Cost of Ownership (TCO) with business drivers, such as enterprise efficiencies, productivity enhancements, and customer satisfaction.

When an organization struggles to lower its total cost of ownership, however, it must first devise a cohesive enterprise strategy meeting daunting challenges such as business units with their own technology standards, applications, and tools; products that do not work effectively together or with existing data; excessive server licenses; and cost-prohibitive support requirements.

Factors That Affect TCO

It is risky to focus solely on price when evaluating costs and reductions. The methodology behind BI's total cost of ownership is multi-faceted as many factors contribute to determining the total cost of ownership of an enterprise BI strategy. Each of these factors is intertwined and dependent upon the others. One should not be removed without considering how it could affect the others. A BI strategy should:

- ■ Support all user needs.
- ■ Allow for the development of custom BI applications.
- ■ Work with your data.
- ■ Run on your platforms.
- ■ Require minimal support.

The costs associated with supporting and providing each metric could become exorbitant as your business grows or the demand for more information increases.

11

CONCLUSIONS

When considering Business Intelligence (BI) software, it is essential to be aware of underlying issues that can seriously affect your bottom line. In a country where most businesses have information stored in several different databases, it's no secret that the BI strategy you choose will affect your profit and loss target. From data collection and integration to data analysis, the holy grail of business intelligence software combines your disparate databases to provide a decision support system that will enable you to make key conclusions using all of the available information. This is clearly a huge bonus to those companies searching for ways to surpass their competitors. However, in today's landscape, managers and executives in small to midsize organizations are at a significant disadvantage where business intelligence is concerned. Their business intelligence data and needs are similar to their Fortune 500 and Global 9000 counterparts, but their business intelligence budgets are not. So how do you find the right business intelligence vendor, one that provides concurrent access to all available types of information with a simple, user-friendly interface at a cost that makes sense? Simple: avoid these top five common obstacles and purchase only what your company will use.

USABILITY VERSUS FEATURE BLOAT

We've all experienced it: we want only the best: the most technically sound flat-screen TV on the market; the newest cell phone with the most ring tones, Web access, and e-mail; a membership to the most popular gym filled with the most technically advanced machines and highly educated trainers. But does too much of a good thing become more overwhelming than beneficial?

Simplify. Are all the bells and whistles necessary? Be selective about the BI programs you choose and only buy what you'll use. Carefully consider the expensive BI products and do your research. There are companies in the BI marketplace that provide decision support systems tailored for the small to midsize company, without the unnecessary expensive add-ons. With a little legwork, you can identify BI software that enables users to integrate, publish, and analyze enterprise data across disparate databases, without expensive ETL (Extraction, Transformation, and Load) technology.

ENOUGH ALREADY ABOUT METADATA!

Metadata is literally data about data. It can range from timestamps to icons to free-text comments and much more. Because of the need to distill the useful information from the mass of information available, it's typical to invest in a large BI company that ensures consistency by manually creating your metadata, which in turn can add value, but also adds tremendous cost. However, it is often possible to leverage companies' existing metadata, a fact that many big BI companies overlook. As such, it may not be necessary to spend the extra money to manually create these metadata schemes; the information you need may already be there. As with anything else, the key is about finding balance: too much or too little data is not useful. Good business intelligence means balanced information.

CONSULTING

Don't buy it if you don't need it! Many BI products require substantial consulting projects as part of their launch process, often costing as much as the software itself. Take a close look at your company's needs and determine whether this additional support is necessary. Unlike the business intelligence products of Big BI, there are niche BI companies with software that installs and runs without expensive consulting, yet they still provide concurrent access to any combination of enterprise databases in a single user request; can be installed in a matter of hours; and include user-friendly tools for query creation, reporting, and data analytics.

LICENSING, UPGRADES, AND MAINTENANCE

Traditionally, business intelligence software requires upgrade and maintenance fees (for support, upgrades, improvements, etc.) beyond the initial expense for the license. These are billed as a function of the "list" license fee. However, that list price can change dynamically as your organization changes the number of user licenses, and the number and speed of processors on the server. Before you know it, your overall "maintenance" costs can increase exponentially. Alternatively, a little research will show that there are companies out there that only charge a flat fee per year—no maintenance fee increases—regardless of how you change your infrastructure. By adopting a decision support system with a flat-fee model, you save money, time, and other resources as well. Avoid the situation where your BI choice has become so big that it's controlling you and you're not controlling it.

RESUME BUILDING

Beware of business intelligence recommendations made for the wrong reasons. Choosing a BI company by its popularity, ubiquity in the marketplace, or by its clientele may mean choosing a company that does not align with your needs, goals, or budget. Although the name may look impressive on paper, the cost may outweigh actual benefits. A little more legwork will unearth BI companies that will provide BI enterprise power without the enterprise cost.

With the proper business intelligence software in place, you can finally move beyond viewing "customers," "workforce," "supply chain," or "finances" (for example) as separate entities, and begin considering all the core elements of your business as a single integrated whole. The result is a system that enables you to measure the accuracy and success of your goals and objectives from various perspectives, and make intelligent decisions based on quantifiable analytics. For these reasons and more, making the right BI choice is key to keeping your P&L on track. By avoiding the five pitfalls mentioned above and keeping your company's needs and objectives top of mind, you'll be well on your way to finding a decision support system that is the perfect fit.

Appendix A

SERVICE-ORIENTED ARCHITECTURES

Service-Oriented Architectures (SOAs) are rapidly gaining importance with every passing day, becoming the buzzword among enterprise technologists. From a business perspective, a service-oriented architecture affords an organization the flexibility to adapt as the business grows and as processes mature and evolve. At a technical level, an SOA provides a standard programming model that allows components to be published, discovered, and invoked over a network. This model provides the infrastructure for an organization's systems to be agile in responding to changing business conditions, yet allows technical implementation details to remain transparent.

SERVICE: THE CORE OF SOA

The central concept in an SOA is the service. A service is a core piece of exposed business functionality that is protocol independent, location agnostic, transport independent, and contains no user state. Services don't contain presentation logic, nor do they contain logic to integrate with back-end systems. Services focus exclusively on solving business domain problems. Some newer implementations of SOAs also define services that can perform technical utility operations. This approach helps an infrastructure support to perform the consistent and reliable execution of common tasks in a constant manner.

Services are very loosely coupled with other components of an application. They're protocol independent, allowing for the same service to be accessed in multiple ways, and coarse-grained, allowing the service to perform its business logic and return the result in a single call. Services typically rely on configuration data to determine things, such as which integration module to call. But configuration data for individual users is typically never stored.

BENEFITS OF SOA

Today's dynamic environment forces organizations to evolve in real-time around their changing business processes. Systems built around service-oriented architectures can adapt and scale rapidly to support the business by leveraging services that adhere to the principles of an SOA. These principles provide benefits that are both business and technical in nature. An SOA provides a range of both business and technical benefits for the corporation.

Business Benefits

Leverage Existing Investments

Organizations have spent time and money developing their existing infra-structure. Today's dynamic environment demands new uses of systems and processes information to help enable more efficient processing and cost-effective business operations. Repurposing systems, processes, and data from existing systems provides one avenue for managing the cost. Exposing legacy systems as services also creates an environment that shares information and processes from systems that were previously disparate or connected through point solutions.

Faster Time to Market

An SOA inherently promotes reuse as part of its core philosophical approach to development and integration. Reuse of services and components allows new applications to be quickly assembled to respond to changing market conditions or business demand.

Risk Mitigation

Risk is defined by the Software Engineering Institute (SEI) as the possibility of suffering a diminished level of success within a software development program. SOA-based efforts reuse core services and processes that have been developed and tested and are well understood, increasing the level of project success of an implementation by reducing potential bug introduction. In addition, once the core architecture has been implemented, development occurs at two levels. One level focuses on orchestration, the assembling of services into processes, workflows, or applications, which requires a focused skill set that is more business-oriented. A second level is more technical in nature and involves the design and development of the services and underlying infrastructure. This separation of skills improves risk management from an organizational perspective by allowing effective allocation of resources to efficiently deliver development tasks.

Continuous Improvement

Service- and component-based development provide continuous improvement because systems and processes communicate via services and interfaces, the implementation of which is encapsulated and invisible to the requestor. Hiding the underlying implementation provides an opportunity for continually improving and optimizing the underlying code base without affecting the use of the service.

Technical Benefits

Location Transparency

Decoupling the client requesting a service and the service itself is known as location transparency. The SOA allows a request to not know (or care) where a component or service is located because of the publication and discovery mechanisms. This eliminates locating, understanding, configuring, and incorporating remote functionality from a development standpoint. The SOA framework should handle all of the publication and discovery work required to use a service. Location transparency is facilitated in an SOA by support for standards such as UDDI in the Web services world.

Loosely Coupled

Coupling refers to the dependencies that exist between software components. When discussing coupling, two main definitions are covered: tight coupling and loose coupling. Loosely coupled implies that decisions can be made by services, components, and applications at runtime rather than compile-time. Loosely coupled components can act independently of each other whereas a tightly coupled approach requires components and dependent components to be available for binding at compile-time as well as runtime. An SOA promotes the design and implementation of loosely coupled applications or platforms by providing mechanisms for supporting loose coupling, and by ensuring that every service is decoupled in time, protocol, and location.

Late Binding

Late binding allows a loosely coupled application to be flexible because there is no inherent knowledge of how the application will be used. It refers to components that determine behaviors and relationships at runtime, rather than at compile- or deploy-time. This dynamism allows applications the flexibility to adjust to changing requests and responses while interactions are occurring. This late binding paradigm allows systems to be reusable, extensible, self-assembling, self-healing, and more maintainable.

Protocol and Device Independence

Services are defined in a manner that is independent of device (either for distribution or presentation), connection mechanism, or transportation protocol. This independence provides the SOA an opportunity to present information to respond to a request from mobile users, desktop users, applications with high or low bandwidth, legacy systems, and so on. Allowing a network to perform in a heterogeneous fashion lets a service "be serviced" by the appropriate system regardless of the network on which it resides.

Service-Oriented Integration Challenges

Here's how the edge-integration scenario invokes service-oriented techniques to enable real-time inter-enterprise collaboration with no manual intervention: An incoming XML or EDI document arrives via a VAN or Internet EDI connection. It is then converted into the company's canonical XML format at the edge of the enterprise for use as part of a business process that gets routed to multiple systems in a service-oriented architecture. This XML document is analogous to the manila folder that preceded electronic information sharing, which contained information defining the transaction and all of the information surrounding the transaction (e.g., approvals, contracts, etc.). The XML business document is continuously changed as it gathers the information from different enterprise systems needed to complete the transaction with the partner. When the process is complete, the document is converted into outbound XML or EDI and returned to the partner, letting it know if the transaction is complete (or can't be completed for whatever reason) or that the next step in the transaction can begin.

This edge-integration scenario is an excellent way to maximize existing technology assets by allowing information to be shared with any XML-enabled application in the enterprise without worrying about breaking the other end of the pipe. Companies also can preserve existing traditional

EDI connections over a value-added network that makes sense while expanding their universe of buyers, suppliers, and partners via Internet AS/2, or native XML using AS/2, SOAP, ebXML, or other transport mechanisms. Perhaps the biggest benefit comes within the enterprise, where the time and cost of processing the transaction has been completely automated using a service-oriented architecture.

In the Web services scenario, a company can make any system available as a service in a secure directory for consumption by a partner or customer. A partner's application will then discover the service and use it to trigger a process that consummates a transaction. The entire process, from publication to discovery to execution, runs in real-time.

In order to successfully implement service-oriented integration techniques and not fall prey to some of the same problems that constrained first-generation BI solutions, companies must meet a number of challenges.

The biggest challenge is enterprise connectivity. Service enabling existing operational systems by hand to support composite and STP applications will undoubtedly result in high cost and high failure rates. Custom service enabling of existing applications has several drawbacks. The first drawback is that the work will only pertain to a single transaction. Although this represents a major improvement over point-to-point integration, because the service can be reused by many different applications, it still necessitates writing new code each time there are new requirements from an existing system. This is a costly and risky way to achieve service-oriented integration.

The second drawback is that custom integrations are extremely hard to maintain. This approach also requires maintaining a competency for each of the back-end systems that need to be tapped. A much more effective alternative calls for deploying an adapter framework that creates a single user interface for accessing any operational system. The major benefit of this approach is that it creates a single connection to each underlying system that can be used over and over again for any initiative. This eliminates the need to create a custom connection for each transaction. Once a system is turned on, it stays on.

Another key challenge revolves around heterogeneous platform environments. Many organizations will have both .NET and J2EE-based platforms. Although Web services theoretically offer a recipe for integration, there are still proprietary elements inherent in the different platforms that make interoperability challenging. By adopting an adapter layer on the middle tier, companies get true any-to-any connectivity. For example, a J2EE developer can access a .NET object, a Web service created in .NET, or a packaged application using the same tooling. And this tooling is equally accessible from any platform's development environment.

A third challenge to service-oriented integration isn't technical but cultural. Many companies have delayed service-oriented projects while waiting to make a major integration platform decision. In doing so, these companies are forgoing the real-time business process improvements of composite applications in the near term. Real-time integration does not have to be a big bang. In fact, moving directly to straight-through processing without building any foundation will result in projects of extremely wide scope and complexity. Companies should get their feet wet with more tactical solutions, provided that these applications are fully reusable by whatever integration platform is ultimately chosen.

REAL-TIME PROJECTS VIA SERVICE-ORIENTED INTEGRATION

The following are recommendations for accelerating real-time projects via service-oriented integration.

Avoid hand-coded integrations at all cost. Create a universal adapter and transformation layer that can service enable all operational systems using a single interface. This eliminates risk, shortens project time, and ensures reusability.

Make sure that integrations created by your adaptation and transformation can plug into any leading integration platform IDE, and that it facilitates interoperability across platforms.

Make sure that the vendor that provides the adapter layer can support all of the projects that you are planning, both inside the enterprise and beyond.

Fill in the white spaces gradually. Don't think that you will dramatically reshape your enterprise in one stroke of a brush. Build incrementally instead. Deliver a series of composite applications, and then reuse them for STP applications. This will greatly decrease the cost of building the RTE by creating near-term ROI that can fund successive applications.

Appendix B

THE ENTERPRISE DATA MANAGEMENT MATURITY MODEL

Achieving the highest level of data management is continuously evolutionary. A company that has not concentrated on the quality of its data cannot expect to progress to the latter stages immediately, primarily because any improvement in data management involves a number of factors and is a gradual process that takes time to show results. To improve the data management level, organizations need to change the entire culture of the organization, from personnel to technology to management strategies.

Understanding the Enterprise Data Management Maturity Model is the first step for any organization looking to improve overall data quality. This allows you to understand where your organization fits within the model and determine what, if any, measures should be taken to advance to the next stage. From there, you can understand when it is appropriate to take on additional responsibilities, processes, and technology to advance further through the model.

The Enterprise Data Management Maturity Model presented here is a tool to educate organizations on how to maximize the value of their data and start treating data as the important strategic asset that it is, or can be.

STAGE 1: UNAWARE

At the initial stage of the Enterprise Data Management Maturity Model, an organization has few defined rules and policies regarding data management. The same or similar data may exist in multiple files and databases. And redundant data could be available in different data sources with different formats and different names. At stage 1, confusion reigns, and cooperation about data issues between departments or job functions is rare.

Companies in the "unaware" category (see Table B.1) have little or no corporate visibility into data management costs or performance. As a result, data quality varies widely across the enterprise. In addition, data management activities are unorganized, and there is no understanding of why problems exist or what impact these problems may have. Data quality "denial" almost certainly exists at this stage. Surprisingly, more than one-third of all organizations are measured at this stage.

Table B.1 Characteristics of an "Unaware" Company

People	*Process*
• Success depends on the competence of a few talented individuals • Organization relies on personnel who may follow different paths within each effort to reconcile and correct data • No management input or buy-in on data integrity problems • Executives do not comprehend the extent of data problems • Organizations tend to blame IT for data quality issues	• No defined data management processes in place. Data management is chaotic and project-focused • "Fire fighting mode." Address problems as they occur through manually driven processes • Infrequent long-range resolution to problems • Redundant data exists throughout the organization, leading to wasted resources across functional units
Technology	**Risk and Reward**
• Tools tend to be general-purpose software (Microsoft Excel, Microsoft Access) and no intensive data management software is in use • No data profiling, analysis, or auditing is used to determine data characteristics • Data cleansing or standardization may occur in isolated areas or data sources • Technologies in place support manual quality improvement methods	• Risk: Extremely high. Data problems can result in lost customers (due to poor understanding of the customer's value) or improper business procedures. A few scapegoats receive the blame, although processes are not in place to properly assign culpability • Reward: Low. Outside the success of an individual employee or department, companies reap very little benefits from data management

Actions Necessary to Advance to Stage 2: Reactive

Due to the risks of the first stage, competitive pressures often serve as the driver for improving data maturity. To progress, companies need to put measures and processes in place to recognize problems with data integrity or usability. Often, it is enough to merely acknowledge that data management issues cause organizational problems and to target the source of these problems. Recognition—plus a commitment to fix data management issues—will help an organization begin to understand data management problems, risks and returns.

STAGE 2: REACTIVE

When an organization reaches stage 2 (see Table B.2), it understands data management problems as they occur. And the organization comprehends that data is critical to its success. Data quality issues are addressed only as major problems occur or projects start to derail. At best, the organization hopes to react to problems to mitigate the severity of outcomes.

At this stage, nonintegrated point solutions perform different specific tasks. Organizations experience variable quality and some predictability in data integrity. In addition, successful individuals receive assignments to the most critical business initiatives to reduce risks and improve results in specific processes. Organizations realize that data management may be of value but are not willing to provide the time and money to prevent problems. Studies show that the largest share of all organizations, 45 percent, fall into the reactive stage.

Actions Necessary to Advance to Stage 3: Proactive

At stage 2, solutions are nonintegrated, disparate point solutions. The impetus for progressing to stage 3 is often a strategic vision by certain managers or executives that better data management processes can lead to tangible business results. To advance, companies have to integrate processes and technologies to achieve more from their data resources. Organizations must also begin to document, establish, and enforce data management policies as a core competence of application development. To ensure that the policies are in place, some level of compliance testing is necessary. And finally, organizations must reach a consensus on ownership of the data management processes and assign responsibility and support.

Table B.2   Characteristics of a "Reactive" Company

People	*Process*
• Success depends on the skills of a group of technical employees (database administrators, IT staff, etc.) • Individuals create useful processes, but no standard procedures exist across groups or locations • Long-range solutions are infrequent • Little corporate management buy-in to the value of data	• Stronger data management roles emerge, but the emphasis remains on correcting data quality issues as they occur • Most processes are short-range and focus on recently discovered problems • Within individual groups and departments, tasks and roles are standardized
Technology	**Risk and Reward**
• Tactical data management tools are often available (such as solutions for data profiling or data quality) • Most data is not integrated, but some individuals or departments attempt integration efforts in isolated environments • Some database administration tactics emerge, such as reactive performance monitoring • Attempts to consolidate data (such as a data warehouse) requires scrap and rework due to data quality issues	• Risk: High, due to a lack of data integration and overall inaccuracy of data throughout the enterprise. Although data is analyzed and corrected sporadically, data failures can still occur on a cross-functional basis • Reward: Limited and mostly anecdotal. Most ROI arrives via individual processes or individuals. Little to no corporatewide recognition of data management benefits

STAGE 3: PROACTIVE

Reaching the third stage (see Table B.3) of the maturity model gives companies the ability to avoid risk and reduce uncertainty. At this stage, data management starts to play a critical role within an organization, as data goes from being an undervalued commodity to an asset that can be used to help organizations make better decisions. As a company in this stage matures, it receives more tangible value from consistent, accurate, and reliable data.

At stage 3, a company looks "beyond the horizon" to understand the impact of data problems on mission-critical information. The requisite technology to support high levels of data inspection and correction are in place. And the organization begins to receive executive- and management-level approval for data management projects.

Table B.3 Characteristics of a "Proactive" Company

People	*Process*
• Management understands and appreciates the role of data management in corporate initiatives • Data management initiatives receive the personnel and resources necessary to create high-quality data • All or most areas of the organization are involved with data management processes • Executive-level decision makers begin to view data as a strategic asset	• Corporate data is more standardized, consistent, and measurable. And preventive measures are in place to assure high levels of data quality • Data metrics are sometimes measured against industry standards to provide insight into areas needing improvement • While in this stage, data management goals shift from problem correction to problem prevention
Technology	**Risk and Reward**
• Data management technology providers become strategic partners with the organization and help define best practices while implementing the technology • A corporate data management group emerges to maintain corporate data definitions, synonyms, business rules, and business value for data elements • Ongoing data audits and data monitoring help the company maintain data integrity over time	• Risks: Medium to low. Risks are reduced by providing better information to increase the reliability of decision making • Reward: Medium to high. Data quality improves, often in certain functional areas and then in broader realms as more employees join the early adopters

Actions Necessary to Advance to Stage 4: Predictive

Advancing to the final stage is as much an evolution of culture as it is an evolution of people, process, or technology. The culture shift starts to change people's behavior, whereas new and better processes and technologies give them a better framework for data improvement.

The advances made in the previous stages provide a solid foundation for data management. To evolve to stage 4, you must implement these advances continuously and consistently, primarily by documenting and replicating best practices throughout the enterprise to reach the pinnacle of the Enterprise Data Management Maturity Model.

STAGE 4: PREDICTIVE

At stage 4, organizations achieve almost complete certainty of results (see Table B.4). Data quality is an integral part of all business processes, and it is engrained throughout the enterprise. Processes are entirely or almost entirely automated. To keep data within accepted limits, data management processes are implemented in real-time and validated continuously.

Because historical issues of data quality are known and understood, data defect prevention is the primary focus of stage 4 organizations. And there are cross-organizational approaches to data quality, helping companies address data problems that overlap business silos. Finally, an important

Table B.4   Characteristics of a "Predictive" Company

People	*Process*
• Full management buy-in for data management processes and standards • Data quality improvement has executive-level sponsorship with direct CEO support • A data management group operates across the organization and has the support of data quality stewards, application developers, and database administrators • Entire organization is committed to "zero defect" policies for data collection and management	• Procedures help the organization achieve the highest levels of data integrity • Processes are in place to ensure that data remains consistent, accurate, and reliable over time through regular monitoring of data quality • New initiatives begin only after careful consideration of how the initiatives will affect the existing data management infrastructure
Technology	**Risk and Reward**
• Data management tools are standardized across the organization • All aspects of the organizations utilize the standard metadata and rules definitions created and maintained by the data management group • Results of data quality audits are continuously inspected, and any variations are resolved immediately • Data models capture the business meaning and technical details of all corporate data elements	• Risk: Low. Data is uniform and tightly controlled, allowing the organization to maintain high-quality information about its customers, prospects, inventory, and products • Rewards: High. Solid, corporatewide data management practices can lead to a better understanding about an organization's current business landscape, allowing management to have full confidence in any data-based decisions

distinction of organizations in this stage is that data management becomes a business process and not a technological tool.

At the final stage of the maturity model, a major culture shift has occurred within the entire organization. Instead of ignoring the implications of data management—or treating data quality as a series of tactical projects—a comprehensive enterprisewide program elevates the process of managing business-critical data. With backing from executive management and buy-in from all business functions, the program can flourish, creating more consistent, accurate, and reliable information to support the entire organization.

GLOSSARY

Aggregated Data Data that is precalculated and stored as a summary in the data warehouse to improve query performance. Example: for every product there might be an aggregate value that details the total number of pieces sold per store per month.

Analytical Applications Software that provides a predefined set of analytical processes and reports targeted at specific horizontal tasks (budgeting) or specific vertical industries (insurance) that leverage a data analysis platform (warehouse, mart, multidimensional database).

Business Intelligence Tools An array of software products that provide users with the ability to access and manipulate (analyze) data. These products may be used as stand-alone or part of a data warehouse solution. Business intelligence software spans simple to complex analysis needs from query and reporting to enterprise reporting to OLAP to analytical application to data mining.

Data Extraction The process of copying (extracting) data from a legacy or production system in order to load it into a warehouse.

Data Mart Smaller data warehouses that contain a subset of enterprise data typically defined along departmental lines. This selectivity of information results in enhanced query performance and manageability of the data. Marts can be dependent (i.e., are fed data from a data warehouse) or independent (i.e., stand-alone).

Data Mining Sophisticated software used to discover meaningful relationships, patterns, and trends from data. This is accomplished by examining the data at its most atomic level (unaggregated) and using pattern recognition technologies, statistical algorithms, and mathematical modeling techniques.

Data Modeling The process of changing the format of production data to make it usable for business reporting.

Data Quality/Scrubbing The processes of analyzing, purging, changing, and reengineering data from its original form into a format suitable for business analysis.

Data Transformation Performed on data after it is extracted from operational sources, including reformatting data, integrating dissimilar data types, and performing calculations.

Data Warehouse A staging area for corporate data, optimized to support information access without affecting production systems. Data from multiple operational sources is extracted at prescheduled times, transformed into one standard format, and loaded into a new database (the warehouse) that is easier to access.

Drill Down The process of navigating from a top-level view down through to the individual detail level. This is a more intuitive way to obtain information.

DSS (Decision Support Systems) Business intelligence tools used together as a system that supports the decision-making processes of a company.

Enterprise Reporting Systems that consolidate disparate application reporting tools into a central report library that creates, stores, manages, and distributes reports across an enterprise to both local and remote users.

Executive Information Systems Business intelligence tools that are aimed at less sophisticated users, who want intuitive (no training required) interfaces to look at complex information.

Metadata Data about data; metadata provides systems with the information needed to properly store and manipulate data.

OLAP (OnLine Analytical Processing) Software that provides users with the ability to examine data within a multidimensional view. So, rather than analyzing data in the traditional two-dimensional manner (product sales by region), users can examine data along three dimensions (product sales by region by time period). OLAP systems are most commonly used for sales forecasting, market trends analysis, and the like.

OLTP (OnLine Transactional Processing) Systems that capture and manage data associated with a company's day-to-day operational processing (order entry, invoicing, general ledger, etc.).

Orchestration Orchestration is the process of composing services into loosely coupled work flows.

Queries Native database commands, usually SQL, used to extract information from a database server. Queries can either browse the contents of a single table or use the database's SQL engine to perform joint conditioned queries, that produce result sets involving data from multiple tables.

Query and Reporting Software Software that enables users to access information contained in database servers and view the results in a report format. Standard reports involve predefined queries that the user cannot change, whereas ad hoc queries allow the user to create queries.

Star Schema A relational database design used to model multidimensional data and enhance performance by creating a central index (fact table).

UDDI Universal Description, Discovery, and Integration or UDDI, is an XML-based catalog or repository of businesses and the Web-accessible services they provide.

REFERENCES

1. Beer, Stafford. *Brain of the Firm.* 2nd edition, Chichester: John Wiley, 1972.
2. Boshyk, Yury: Beyond knowledge management: How companies mobilize experience. In *Mastering Information Management,* ed. Donald A. Marchand, and Thomas H. Davenport. London: Prentice-Hall, pp. 51–58, 2000.
3. Collins, B. *Better Business Intelligence — How to Learn More about Your Company.* Letchworth: Astron On-Line, p. 164, 1997.
4. Cortada, James W. and Woods, John A. *The Knowledge Management Yearbook 2000–2001.* Boston: Butterworth-Heinemann, 2000.
5. Darbe, A. Business intelligence is more than a buzzword, *Credit Union Magazine,* October, 33–34, 2001.
6. Davenport, Thomas H. and Marchand, Donald A. Is KM just good information management? In *Mastering Information Management,* ed. Donald A. Marchand, and Thomas H. Davenport. London: Prentice Hall, pp. 165–169, 2000.
7. Davenport, Tometal. Data to knowledge to results: Building an analytical capability. *California Management Review,* pp. 117–138, 2001.
8. Dekker, Hans. *Business Intelligence,* Emerce, 2002.
9. Forsman, Sarah. OLAP Council White Paper, OLAP Council, 1997.
10. Gorinson, Stanley M. and Falis, Neil D. Accounting, patriot laws raise concern for telcos — Public and private. 28 April 2003. *Phone Plus Magazine* Web site, 2003.
11. Gueldenberg, S. C. Measuring in the knowledge age: The perspective of the living and learning organization. *Journal of Strategic Performance Measurement,* pp. 6–15, 1999.
12. Halliman, C. *Business Intelligence Using Smart Techniques.* Houston: Information Uncover. p. 212, 2001.
13. Harryson, S.J.: *Managing Know-Who Based Companies.* Northampton: Edward Elgar, 2000.
14. Kalakota R. and Robinson, M. *e-Business 2.0 – Roadmap for Success.* Boston: Addison-Wesley, 2000.
15. Kaplan, R. and Norton, D. The balanced scorecard—Measures that drive performance. *Harvard Business Review,* Jan.–Feb. 1992.
16. Kaplan, R. and Norton, D. Putting the balanced scorecard to work. *Harvard Business Review,* p. 128, 1983.
17. Kaplan, R.S., Norton, D.P. *The Balanced Scorecard. Translating Strategy into Action.* Boston: Harvard Business School Press, 1996.
18. Leonard-Barton, Dorothy. *Wellsprings of Knowledge. Building and Sustaining the Sources of Innovation.* Boston: Harvard Business School Press, 1995.

19. Lynch, R.L. and Cross, K.F. *Measure Up! The Essential Guide to Measuring Business Performance*. Mandarin, 1991.
20. Marchand, Donald A. and Davenport, Thomas H. (eds.) *Mastering Information Management*. London: Prentice Hall, 2000.
21. McKnight, William. Business intelligence in enterprise portals. *DM Review*. December, 2002.
22. Neely, A. *Measuring Business Performance. Why, What and How?* London: Profile, 1998.
23. Neely, A. and Adams, C. *Perspectives on Performance: The Performance Prism*. Centre for Business Performance, Cranfield, UK: Cranfield Business School, 2000.
24. Neely, A., Mills, J., Gregory, M., Richards, H., Platts, K., and Bourne, M. *Getting the Measure of Your Business*. London: Findlay, 1996.
25. Neely, A., Mills, J., Platts, K., Richards, H., Gregory, M., Bourne, M., and Kennerley, M. Performance measurement system design: Developing and testing a process-based approach. *International Journal of Operations & Production Management*, 20:10, 2000.
26. Nonaka, Ikujiro and Takeuchi, Hirotaka. *The Knowledge Creating Company. How Japanese Companies Create the Dynamics of Innovation*. New York: Oxford University Press, 1995.
27. O'Connor, Flannery. Everything that rises must converge. In *Flannery O'Connor: The Complete Short Stories,* ed. Robert Giroux, New York: Farrar, Straus, and Giroux, 1985.
28. *PMA (Perspectives on Performance) Newsletter,* 1:4 (Oct.), 2001.
29. Porter M. The value chain and competitive advantage. In *Competitive Advantage: Creating & Sustaining Superior Performance*. New York: Free Press, 1985.
30. Simons, R. *Performance Measurement & Control Systems for Implementing Strategy*. Upper Saddle River, NJ: Prentice Hall, 2000.
31. Viva Business Intelligence. Introduction to business intelligence. Pro-How Paper 1/98, 1998.
32. Wah, Luisa: Behind the buzz: The substance of knowledge management. In *The Knowledge Management Yearbook 2000–2001*, ed. James W. Cortada and John A. Woods. Boston: Butterworth-Heinemann, 2000.
33. Watson, Hugh, Houdeshel, George, and Rainer, Rex. *Building Executive Information Systems*. New York: Wiley, 1997.

SOURCES

Papers and Articles

1. Architecting for Agility Source: Gartner Group
   http://www4.gartner.com/DisplayDocument?doc_cd=110392
2. Dawn of the real-time enterprise Source: InfoWorld
   http://www.infoworld.com/article/02/01/17/020121fetca_1.html

3. Managers Need Real-Time Initiatives for Strategic Decisions Source: Gartner Group
   http://www4.gartner.com/DisplayDocument?doc_cd=110367
4. Now is the Time for Real-Time Enterprise Source: Gartner Group
   http://www4.gartner.com/pages/story.php.id.2646.s.8.jps
5. The Real-Time Enterprise Puts Customers in the Driver's Seat Source: DestinationCRM
   http://www.destinationcrm.com/print/default.asp?ArticleID = 2884
6. Start Planning Now for the Real-Time Enterprise Source: Gartner Group
   http://www4.gartner.com/DisplayDocument?doc_cd=110483

Company Home Pages

1. Brio Technology
   http://www.brio.com
2. BI Technologies
   http://www.businessintelligencetechnologies.com
3. Business Objects
   http://www.businessobjects.com
4. Cognos
   http://www.cognos.com
5. CorVu
   http://www.corvu.com
6. Crystal Decisions
   http://www.crystaldecisions.com
7. Gartner Group
   http://www.gartner.com
8. Hummingbird
   http://www.hummingbird.com
9. Hyperion
   http://www.hyperion.com
10. IBM
    http://www.ibm.com
11. Microsoft
    http://www.microsoft.com
12. MicroStrategy
    http://www.microstrategy.com
13. OLAP Council
    http://www.olapcouncil.org
14. OLAP Solutions
    http://www.olapsolutions.co.uk

15. Oracle
    http://www.oracle.com
16. SAS Institute
    http://www.sas.com

Journals and Magazines

1. Database Magazine
   http://www.array.nl/dbm
2. Datawarehouse Infocenter
   http://www.dwinfocenter.org
3. Intelligent Enterprise
   http://www.intelligententerprise.com
4. KDnuggets
   http://www.kdnuggets.com/publications/index.html

INDEX